D0873422

EINSTEIN'S
GOD

EINSTEIN'S GOD

Albert Einstein's Quest as a Scientist and as a Jew to Replace a Forsaken God

ROBERT N. GOLDMAN

JASON ARONSON INC.
Northvale, New Jersey
London

Excerpts from Albert Einstein's archival materials reprinted by permission of the Albert Einstein Archives, The Hebrew University of Jerusalem, Israel.

"Time's Eye," and "A Rumbling," from *Poems of Paul Celan*, translated by Michael Hamburger. © 1972, 1980, 1988, 1995 by Michael Hamburger. Reprinted by permission of Persea Books, Inc.

Lichtenberg: A Doctrine of Scattered Occasions, J. P. Stern (Indiana University Press, 1959). © J. P. Stern. Reprinted by permission of author.

Aphorisms, by Georg Cristoph Lichtenberg, translated by R. J. Hollingdale (Penguin Classics, 1990). Trans. © 1990 R. J. Hillingdale. Reprinted by permission of Penguin Books, Ltd.

This book was set in 12 pt. Antiqua by Alpha Graphics of Pittsfield, New Hampshire.

Library of Congress Cataloging-in-Publication Data
Goldman, Robert N.
    Einstein's God : Albert Einstein's quest as a scientist and as a
Jew to replace a forsaken God / Robert N. Goldman.
        p.   cm.
    Includes bibliographical references and index.
    ISBN 1-56821-983-0 (alk. paper)
    1. Einstein, Albert, 1879–1955—Views on God.   2. God (Judaism)
3. Philosophy and religion.   I. Einstein, Albert, 1879–1955.   II. Title.
QC16.E5G67   1997
211'.092—dc20   96-33633

Manufactured in the United States of America. Jason Aronson Inc. offers books and cassettes. For information and catalog write to Jason Aronson Inc., 230 Livingston Street, Northvale, New Jersey 07647.

To Judy, Lita, and Erica

The bigotry of the nonbeliever is for me nearly as funny as the bigotry of the believer.

EINSTEIN

Modern thought, Albert Einstein wrote, "*has deprived the more wide-awake intellectual of the feeling of security and of those consolations which traditional religion, founded on authority, offered to earlier generations. . . . Often he escapes from this misery into a fanciedly superior skepticism, or into distractions of all kinds which keep him from coming to his senses. But the effort is in vain. We cannot in the long run substitute narcotics for wholesome nourishment.*"

Contents

Acknowledgments

I am grateful to Harry Woolf, then Director of the Institute for Advanced Study, for inviting me to do research in the Einstein Archives and to Helen Dukas, Einstein's secretary, for her wholehearted cooperation.

I am indebted to many in writing this book. Robert Littman helped from the very beginning with practical advice on navigating the shoal-strewn waters between rough draft and published work. Iris Wiley read several drafts meticulously, offering criticism that could only come from an experienced editor (which she is). Jamie Sayen was doing research in the Einstein archives at the same time as me, and I benefited tremendously from our discussions. Irving Copi, Judith Goldman, and Rabbi Herbert Weiner read early drafts and provided constructive ideas. Rabbi Gunther Plaut and Elizabeth Plaut supplied encouragement and advice at a needed time. I am grateful to Rabbi Plaut for permission to include (in Chapter 3) his own translation into English of a letter to him from Einstein concerning the religious sense. Alexander Hoyt told me how to make the format more commercially palatable. Val Mori translated Einstein's German into accurate but clear English. Erica Goldman and Jordan Popper made valuable suggestions. And, of course, this book could not have been written at all without my wife's patience and understanding during its gestation.

Needless to say, especially since I did not always listen to the excellent advice proffered, I alone am responsible for any errors or infelicities in the text.

Preface

One morning, shortly after accepting an invitation to study Einstein's letters and documents in Princeton, I entered Helen Dukas' office to discuss them with her. She had been his longtime personal secretary, who continued to be in charge of his papers after his death. There I found her talking to Abraham Pais, Einstein's colleague and biographer, who knew him during the last nine years of his life. Pais asked me in what area my interests lay. "Einstein's mysticism . . . ," I began to reply but he interrupted me, laughing. "Einstein was no mystic," he chortled, and then went on to say Einstein was one of the least mystical men he had ever met.

Pais was right. Discomfited, I realized I had used the wrong phrase to describe Einstein's concern with the human condition, with matters of mystery, of spirit, of self-transcendence. For him, mystery is material; spirit is human. Experiencing either can be a religious experience, but neither represents a vague, separated mystical realm.

I later came across a compilation from his writings that had been submitted to him during the last days of his life to edit for publication. One quotation had been neatly typed:

> The most beautiful and most profound emotion that we can experience is the sensation of the mystical. . . .

He had crossed out the word "mystical" with his pen. In its place he wrote, "mysterious."

Einstein, with his earthy sense of humor and carnal appetite, abhorred mysticism. But he read widely, thought deeply, about human existence. His sophistication in such matters philosophical and religious is easily missed because he expressed himself in simple language—often just in letters rather than essays, rarely in learned papers.

He wrote a great many letters, often with an unself-conscious spontaneity that reveals more about his intimate thought, religious and other, than do his formal writings. He wrote to friends— his letters to some span his adult lifetime. He wrote to strangers—because he was kind, he tried to reply to the most worthwhile of the unending stream of letters asking for his opinion or help. To learn about Einstein's ideas and views, I have delved into these letters, most of which are still unpublished.

The letters at the time were held in Princeton at the Institute for Advanced Study. There I met the remarkable Miss Dukas, who has died since, and was fortunate in having her help. She had set up the correspondence files and was very familiar with their contents. Since she had been with Einstein in both Germany and America, she often was able to comment on the background of the letters she so briskly located.

I remember a significant incident the day we met. I had been telling her of my interest in Einstein's fascination with Spinoza, whom he called "the greatest of modern philosophers." (The second quotation on page vii is from Einstein's introduction to a book on Spinoza written by his son-in-law.) We were in her office at the Institute. She immediately went to the nearby files, where Einstein's letters were held, pulled out one, made a Xerox copy of it, and handed the copy to me. From those massive correspondence files and from the many letters mentioning Spinoza, she had unhesitatingly selected one that, though it seemed innocuous to me at the time, I later came to realize was a key document in understanding Einstein's attitude to Spinoza and to life. (The gist of the letter is quoted at the beginning of chapter 10.)

He had willed his papers to the Hebrew University in Jerusalem, where they now are, but with the proviso that they would remain in Princeton under Helen Dukas' care during her lifetime. Meeting her—unpretentious, fiercely loyal to his memory, almost always patient but with a refreshing no-nonsense attitude—was both a delight and a privilege.

Albert Einstein disliked pretension in all its forms, including that in writing. Upon reading Aristotle he protested to a friend, "If it had not been so obscure and so confusing, this kind of philosophy would not have held its own very long. But most men revere words they cannot understand and consider a writer they can understand to be superficial." He wanted books to state their ideas in clear language and to be no longer than necessary. He himself wrote that way. I have tried to write this small book that way.

If I have not succeeded, I will try to take probably unjustified comfort from a question raised by that wily eighteenth-century savant whose comments Einstein savored, Georg Christoph Lichtenberg. It was a remark that Einstein once quoted in referring to himself and a book he was reading. Inquired Lichtenberg: "When a book and a head collide, and there is a hollow sound, is it always the fault of the book?"

1

"I Am a Deeply
Religious Nonbeliever"

*I thank the Lord a thousand times that he let me be-
come an atheist.*

GEORG CHRISTOPH LICHTENBERG

"I have been absent from a synagogue so long that
I am afraid God would not recognize me." With
these words Albert Einstein gently turned down
an invitation from a Florida rabbi to attend ser-
vices with his congregation.

Though he had scant regard for religious ritual,
he was fascinated by the word "God" and used
it often in casual conversation. His close assistant
for several years, Leopold Infeld, who had been
brought up in Poland and so should know, said
Einstein used the word "God" more often than did
a Catholic priest. One of his lines about God has

become a catchphrase: "I do not believe that God plays dice with the universe."

Here he is expressing his belief that there are no chance events in the universe. Most physical scientists today disagree with him. Cambridge University's Stephen Hawking has retorted that God not only plays dice with the universe, he throws the dice where nobody can see them. But it is the truth that is really hidden. Whether it is Einstein or Hawking who is more correct—or even whether it makes any sense to compare the two assertions—remains unknown.

Another of Einstein's statements about God is carved above a fireplace at Princeton University, where it appears in the original German in which Einstein expressed it. Its English translation is, "God is subtle, but he is not unkind." Those were his words when he was told that another scientist's work—the ether drift experiments of D. C. Miller— had disproved his theory of relativity. Notice that Einstein assumes God is on his side. Fortunately for him, the report turned out to be false.

But one of Einstein's more thought-provoking lines about God, which he expressed to Leo Szilard, is known by very few: "As long as you pray to God and ask him for SOMETHING, you are not a religious man."

Not all scientists appreciated Einstein's frequent invoking of God. One, the Russian physicist Peter Kapitza, was irritated by it. Kapitza, a Nobel Prize winner, snapped, "Einstein loved to refer to God when there was no more sensible argument."

But, to give Einstein more credit than did Kapitza, what did he really have in mind when he used the word "God"? Was it merely a picturesque manner of speaking in which he said "God" when he meant "nature"? Or did he mean something more?

He did not mean a personlike God. He likened the reasoning of those who believed in such a God to "the animistic thought of primitive people [that] if there is a wind there must be somebody who is blowing."

In 1940, a symposium calling together scientists, philosophers, and religious leaders was held at the Jewish Theological Seminary in New York. Einstein wrote a paper for it in which he said: ". . . teachers of religion must have the stature to give up the doctrine of a personal God, that is give up that source of fear and hope which in the past placed such vast power in the hands of the priests."

Upon reading his words, the seminary's head, Rabbi Louis Finkelstein, may have regretted inviting him. Finkelstein said he was surprised that "Professor Einstein should give such an absolute judgment in a field that was philosophical and theological in character." Einstein, he said reprovingly, "should realize that he must speak with as much reserve in these fields as he habitually does in his own field of natural science."

Time magazine, with its journalistic instinct to pounce when it discerned any weakness in a public figure, called Einstein's paper "the only false note of the entire conference."

Einstein on a Personal God–from the 1940 Jewish Theological Seminary Symposium

... the doctrine of a personal God interfering with natural events could never be *refuted*, in the real sense, by science, for this doctrine can always take refuge in those domains in which scientific knowledge has not yet been able to set foot.

But I am persuaded that such behavior on the part of the representatives of religion would not only be unworthy but also fatal. For a doctrine which is able to maintain itself not in clear light but only in the dark, will of necessity lose its effect on mankind, with incalculable harm to human progress. In their struggle for the ethical good, teachers of religion must have the stature to give up the doctrine of a personal God, that is, give up that source of fear and hope which in the past placed such vast power in the hands of priests. In their labors they will have to avail themselves of those forces which are capable of cultivating the Good, the True, and the Beautiful in humanity itself. This is, to be sure, a more difficult but an incomparably more worthy task.* After religious teachers accomplish the refining process indicated they will surely recognize with joy that true religion has been ennobled and made more profound by scientific knowledge.

*This thought is convincingly presented in Herbert Samuel's book, *Belief and Action*.

Einstein once declared to a scientist friend, also of Jewish background, "The search for truth is something that can really replace the religion of our fathers." But in an apparently contradictory— though nicely ecumenical—statement, he also once remarked that the achievements of Buddha, Moses, and Jesus are higher than those of the scientists, more important to mankind, calling the three of them "blessed men." He often referred approvingly to the ideas of Gandhi and Spinoza, both of whom proclaimed, "God is Truth," though each meant something quite different by the phrase. And he also wrote:

> [About] the question whether the belief in a personal god should be attacked. Freud in his last writings was of this opinion. I myself would never be part of such an undertaking because such a belief seems to me still better than the lack of any kind of transcendental interpretation of life. And it seems doubtful whether one can offer to most men with any success a sublime means to satisfy the metaphysical need.

> Atheists are creatures who . . . cannot hear the music of the spheres. The wonders of nature do not become smaller because one cannot measure it by the standards of human morals and human aims.

In his last year he said of himself, "I am a deeply religious nonbeliever." Though he rejected the idea of a personal God, he searched for what he believed transcended that idea, drawing from Jewish tradition in doing so.

He enjoyed thinking in biblical imagery, expressing himself in biblical terms. At the end of his life he considered himself a failure because he had not succeeded in developing a unified theory underlying the forces of nature, what some now call a TOE—"theory of everything." We know today that it would have been impossible for him to do so because the scientific data available in his time was insufficient. But he did not know that. In writing to a friend in Israel, he told of his feelings about his work this way:

> So it is imaginable that it is only chimera or fancy. Thus I am even worse off than our Moses because he could at least see before his departure that the promised land was before him while what I am seeing may be chimera. Well, so what—at least it is a beautiful one.

He had not wanted his comments to the Jewish Theological Seminary symposium to be published because he thought they would be misunderstood. He was right; they were. His religious thought was self-contradictory only on the surface. In the pages ahead we will explore it.

A JEWISH CHILD

Einstein's enthrallment with the word "God" and with biblical story dated from childhood.

When he was born in southern Germany, in 1879, the Jews were free from any legal restriction but had not been free long enough to feel fully secure about it. They had been granted citizenship only eight years earlier. Many German Jews were trying to detach themselves from their Jewish background—religious, cultural, or both. Among them was Albert's father, Hermann Einstein.

Hermann Einstein prided himself on being a freethinker and had rejected Jewish religious practice. Young Albert, however, learned Jewish customs and ways from pious grandparents and other family members. When he was an adult, he reported that he had grown up "in a characteristically Jewish environment."

He received instruction in both Bible and Talmud and was a very religious child: thinking much about God, writing little poems in praise of God, singing songs to God. He tried to adhere to Jewish law and asked his parents to observe the Jewish dietary laws at home; but his father refused to make concession to what he called "dogmatic ritual." Albert refused to eat pork anyway.

But it was a Jewish custom that his family did observe—a custom dating from ghetto days—that abruptly ended this enthusiasm for Judaism and for God. Once a week they would invite a poor Jewish student, living away from home and so presumably in need of a good meal, to have dinner with them. For five years, beginning when Albert was ten years old, their guest was a medical student, Max Talmud. Talmud brought young Einstein books on science to read and discussed them with him. The impressionable, logically inclined boy soon became convinced "that much in the stories of the Bible could not be true."

Questioning one's childhood faith is common in adolescence. But young Albert Einstein's feelings were intense. For him it was a major crisis. Looking back upon it sixty years later, he wrote in his scientific autobiography:

> The consequence was a positively fanatic [orgy of] freethinking coupled with the impression that youth is intentionally deceived by the state through lies; it was a crushing impression. Suspicion against every kind of authority grew out of this experience.

He felt betrayed. Disillusioned by religion, yet needing an outlet for the religious fervor bursting out of him, he turned its focus away from God and onto science and philosophy. His quest had begun.

2

To Replace a
Forsaken God

*With most men, unbelief in one thing springs from
blind belief in another.*
GEORG CHRISTOPH LICHTENBERG

He wanted to be a physics teacher. But Albert Einstein was unable to find a job when he graduated from college—even though he had attended the prestigious Federal Institute of Technology in Zurich. He believed that he was discriminated against because he looked Jewish—he thought he had a Jewish nose—and felt sorry for himself.

However, in fairness to the Swiss society of the time, there appeared to be other reasons. Not only did he seem arrogant, but he had not been a good student. He had cut classes often, passing examinations only by studying the lecture notes taken by a classmate, Marcel Grossman. His college girl-friend, Mileva Maric, who was of Greek Orthodox background, wrote to one of her confidants, "Yet, it is not likely that he will soon get a secure position; you know that my sweetheart has a very wicked tongue and is a Jew into the bargain."

He finally did get work—as "patent examiner, third class" with the Swiss patent office—but only through the influence of Marcel Grossman's father, who knew the director of the office. He now married Mileva and also entered upon the most intellectually fruitful period in his life.

In the year 1905 he published a number of papers that influenced the course of physical science. In one, which became known as the special theory of relativity, he changed our understanding of time—and set the stage for changing his own outlook on life.

THE NATURE OF TIME

We tend to think of time as a smooth, riverlike flow in which we all move together from past to future through a changing present moment. No, said Einstein. There is no uniform flow of time in the universe. The flow of time will differ for two bodies if their states of motion sufficiently differ.

The classic example is the story of the twins, one of whom is an astronaut. If that twin were able to leave earth in a rocket that is flying fast enough and long enough he would return to earth noticeably younger than his brother. The astronaut may believe he has been away two years, and he has actually aged two years. But his twin who remained on earth believes that ten years have elapsed and has actually aged ten years. The theme is a familiar one in science fiction. We do not, by the way, know how to build rockets capable of going that fast.

Two events, said Einstein in his 1905 paper, that appear simultaneous to one cosmic observer may to a second observer, moving in relation to the first, appear to be happening at different times. If two events A and B are so widely separated in the universe that neither can affect the other, one observer may see event A happening before event B, while the second observer sees B occurring before A. There is no uniform, flowing present moment to define which is earlier and which is later. The universe has no "now"—each observer provides his own.

Stranger still, from the viewpoint of a moving observer elsewhere in the universe, all the events of your own life—past and future—may appear as a vast panorama in which your life episodes are spatially separated but simultaneous.[1]

––––––––––

1. According to relativity theory, the events observed cannot be precisely simultaneous but can be so close to it that the observer will not be able to tell the difference.

That is, although there must be a tiny time difference between successive frames of observation, that difference can be asymptotically close. Consequently, *all* the frames of being constituting a person's life can be within the persistence of vision representing the "now" of the observer.

Every experiment to test Einstein's theory of special relativity that has been properly performed has confirmed its truth.

HERMANN MINKOWSKI'S BRILLIANT IDEA

Hermann Minkowski, Einstein's former mathematics professor in Zurich, remembered him as a headstrong "lazy dog." And the feeling seems to have been reciprocal: Einstein had advised a schoolmate that it would not be worthwhile to take Minkowski's courses. Despite this mutual lack of appeal, Hermann Minkowski was destined to play a fundamental role in the development of Einstein's concept of space and time.

Minkowski read Einstein's paper on the special theory of relativity with astonishment. He recalled that Einstein had neglected his mathematics courses because he could not see their relevance to the physics problems that fascinated him. In later years he often said to his students: "Einstein's presentation of his deep theory is mathematically awkward—I can say that because he got his mathematical education in Zurich from me."

Einstein's paper stoked the fire of Minkowski's fertile imagination—Minkowski had been a mathematics prodigy as a youngster—and he was hit by a stunning insight: The theory of relativity could be interpreted in terms of four dimensions in which space and time were not detached from each other, but were interrelated; time and space as independent entities do not exist. Each is but an aspect of a more fundamental reality, spacetime, which makes up the structure of the universe.

For example, the time between two events in the universe and the distance separating them are not fixed—they depend on the state of motion of the observer doing the measuring. Observers moving at different speeds would see either the time difference or spatial separation as larger and the other measurement as smaller. It is a combination of the two, called the spacetime interval, which is constant. Minkowski's words, in which he first introduced these ideas in a public lecture, have become famous: "From now on, space by itself and time by itself must sink into the shadows while only a union of the two preserve independence."

To Minkowski, it was a wonderfully elevating revelation. Born of Jewish parents in a rural Polish ghetto, he had received a traditional Jewish education with its emphasis on Talmud and Bible. But he had permitted himself to be baptized so as not to hinder his career. Baptized out of convenience, having lost the strong Jewish belief of his childhood without having found any substitute, he was spiritually adrift.

A mathematician, he sought an answer to his malaise in mathematics—specifically in the abstract beauty of the theory of numbers. The physicist Max Born writes that for Minkowski "number theory was the most wonderful creation of the human mind and spirit, equally a science and the greatest of arts." The theory of numbers is concerned with the relationships between integers (that is, the digits 1, 2, 3, and so on). But to Minkowski's dismay, he found no spiritual satisfaction there because of its lack of relation to human truth and passion.[2]

Now, however, in his vision of the unity of space and time, he believed that he had created the basis for a new view of nature and there found meaning for his life. He played with various names for it, trying out "world surface," "world mirror," and "cosmograph." He finally called it "The Theory of the Absolute World." He had found his substitute for the absolute biblical God of his childhood.

Obsessed with his vision of Truth, Minkowski strove to perfect his theory. But four months after giving his famous public lecture he was stricken with appendicitis. In the hospital, he knew that at age forty-five he was going to die. He decided to use those last hours to correct the proof sheets for his latest work on spacetime. He wanted his Theory of the Absolute World to be better understood.

Einstein—true to his feelings about Minkowski and mathematics as an undergraduate—had at first not been impressed by Minkowski's methods, which, in their succinct mathematical form, are sophisticated and equation-laden. He considered Minkowski's approach to be "superfluous learnedness." But he soon learned to appreciate its power and used it in generalizing his theory of relativity.

---

2. Today the theory of numbers does have relation to human affairs. It provides the basis for a powerful way of encrypting data, "trapdoor codes," used to maintain privacy in communication networks.

The pure unadorned beauty of Minkowski's fundamental concept is caught in Einstein's explanatory words:

> Space and time were merged into a single continuum in the same way as the three dimensions of space had been before. Physical space was thus increased to four-dimensional space which also included the dimension of time. The four-dimensional space of the special theory of relativity is just as rigid and absolute as Newton's space.

WITH THE JEWS OF PRAGUE

Einstein became known in the scientific community after his papers were published, and he was now offered real, though modest, teaching positions. While lecturing at the University of Zurich, he learned that he was being considered for a full professorship outside Switzerland—at the German University of Prague.

He very much wanted to go to Prague and make a fresh start. The salary was higher, the prestige greater, and he already had managed to alienate some members of the faculty in his short stay at Zurich. He seems to have had a talent for discomfiting people whose opinions he did not respect.

Since leaving home he had paid little attention to his Jewish background—except for worrying whether looking Jewish would hurt his chances of getting a job. When his wife, Mileva, had converted from Greek Orthodoxy to Roman Catholicism, he merely commented, "It's all the same to me." Now, waiting impatiently to hear from Prague, he was concerned that the Jewish ancestry he disregarded might prevent his appointment.

The offer did arrive—but the religious difficulty was not the one that Einstein expected. Prague at that time—1911—was part of the Austro-Hungarian empire. Emperor Franz Joseph had stipulated that all teachers in his domain declare some religious affiliation, no matter what it might be. Einstein tried to reply "none," but that answer did not satisfy the emperor's edict.

Though he had rejected religion when he became disillusioned as a teenager, he was now being forced to choose one. Should he stand by his principles, insist that he had no religion, and probably lose the position? Or should he say that he was a Jew? That, he knew, would make further promotion difficult. Or should he, as so many others of Jewish birth had done in order to make career advancement easier, declare himself "Christian"?

But he could not bring himself to write in the word "Christian." Albert Einstein, at the age of thirty-two, pronounced himself a Jew.

It was a symbolic beginning. For, in the city of Prague, Einstein began to return to the Jewish community. It was not voluntary—he was given no choice in the matter. Prague had a tradition-laden Jewish past and a pervasive Jewish present. Together, they manifested themselves so strongly that even a man such as Einstein then was, thinking only of physics and philosophy, could not evade them.

Seventy years before his arrival, Prague's Jews had been granted freedom to move out of the ghetto area to which they had been confined for centuries, but historic structures in the old Jewish quarter had been preserved. Ancient synagogues—the oldest built in the eleventh century—still stood. A clock on the old Jewish town hall proclaimed the time in Hebrew lettering. The old Jewish cemetery, adjoining the ghetto area, contains twelve thousand tombstones, dating from the year 1439, their inscriptions still readable.

Einstein wandered through these ancient monuments to his people and was reminded of his heritage: reminded of the rich Jewish background and the close warm family in which he had grown up.

His appointment was not to the University of Prague but to the *German* University of Prague. There was also a *Czech* University of Prague, larger but less prestigious. Originally the two were a single institution, but the Austrian government had split it into two a quarter-century earlier. It was a wise move because the people of Prague were split into two groups, one that spoke German and one that spoke Czech, and each refused to associate with the other. The Germans were greatly outnumbered by the Czechs but considered themselves more cultured. They were certainly richer.

Over half of the Germans, and a small percentage of the Czechs, were Jews. The non-Jewish Germans and Czechs did have one area of agreement:

they both disliked Jews. The Jews too were divided, albeit more peaceably. There were German Jews, Russian Jews, and Czech Jews in Prague. As the Germans looked down upon the less-cultivated Czechs, the German Jews looked down upon the less-cultivated Russian Jews.

Fate thus decreed that Einstein would become part of the city's Jewish community. He joined a Jewish intellectual circle devoted to philosophy and chamber music that met at the home of a local pharmacist and his wife. The group was reading Hegel when Einstein began to attend.

They were a formidably cultivated assemblage of people—all questioning their religion or lack of it, some seeking to enrich their Judaism, and others, to replace it. There was Hugo Bergman, a talented young scholar. He had once been advised by the prominent philosopher Franz Brentano to convert to Christianity. Only then, said Brentano, could he attain the academic status his abilities merited. There was Max Brod, a prolific novelist who became prominent as the biographer and literary executor for Franz Kafka. And occasionally there was Franz Kafka, who later became famous for being Franz Kafka.

Bergman never followed Brentano's advice, never did convert to Christianity. Instead, he later settled in Jerusalem, where he joined the faculty of the new Hebrew University and became a distinguished professor of philosophy. Brod too emigrated to Palestine in later years. Kafka, on the other hand, rejected his Jewishness and despaired of finding a substitute foundation—a despair that enriched his writing. "What do I have in common with the Jews?" Kafka wrote; "I don't even have anything in common with myself."

Another Jew destined for fame, Martin Buber, then in his early thirties, came to Prague to speak to the Jewish university students about Hasidism and Zionism. While in Prague he sought out Einstein to engage him in dialogue. He wanted to query Einstein about God. Buber reports on their encounter: "I had been pressing him in vain with a concealed question about his faith. Finally he burst forth. 'What we (and by this "we" he meant we physicists) strive for,' he cried, 'is just to draw his lines after *Him*.'"

Buber was not impressed. He thought that Einstein was exhibiting an "innocent hubris" in a questionable attempt to understand the ways of God "as one retraces a geometrical figure."

It is doubtful whether the mystically inclined Buber and the realistic-thinking Einstein would have been able to have a true meeting of minds. Said Kafka of Buber: "No matter what he says, something is missing."

But it is possible that Einstein's relativity provided inspiration for the philosophy of dialogue on which Buber's fame is based. At this time Buber's major philosophical interest was mysticism, and his most recent book had been about Chuang-Tzu and Taoism. He developed his concepts of I–Thou in the years following his visits to Prague—he writes that he planned the book in 1916. In the beginning of *I and Thou*, in the book's original edition, Buber lays down the theme of his new approach to philosophy:

"Man's basic words are not single words but word-pairs. Basic words do not signify things but relations." It would not have been the first time that physical ideas provided a model for psychological concepts.

Einstein's assistant at the university was a young Jewish man, Emil Nohel, the son of a farmer in a nearby village. The professor who succeeded Einstein in his position at Prague, Philip Frank, described Nohel as having "the quiet poise of a peasant rather than the nervous personality so often found among the Jews." Nohel told Einstein about Jewish farmers and tradesmen in the surrounding country and small towns. Living alongside the peasants, they would speak Czech during the week but only German on the Sabbath; they used German as a substitute for the Hebrew they had forgotten.

Nohel was a type of Jew, with rustic rather than urban traits, that Einstein had not met before. He opened Einstein's curious mind to the varied, as well as precarious, nature of the Jewish world.

One week an outstanding young Viennese physicist, Paul Ehrenfest, came to Prague to visit Einstein. Ehrenfest, unable to find a position because of his Jewish background, was looking for a job. Einstein, who was now planning to return to Zurich after little more than a year in Prague, proposed that Ehrenfest take his place. But Ehrenfest refused to do what Einstein had done—declare himself to be a Jew in order to satisfy Emperor Franz Josef's religious affiliation requirement. Nor was he willing to declare himself Christian. He had met his wife, Tatyana, a Russian girl of Christian birth, while mountaineering and had followed her back to Russia to marry her. They had officially declared themselves to be without any religion since it was against the law for Jew and Christian to marry. He now refused to change that declaration in spite of Einstein's assurance that it was only a matter of form.

To Einstein his Jewishness was still a matter of form. But in Prague, with its ancient Jewish presence, the seeds of his awareness of himself as a Jew were planted and nurtured. In time those seeds would sprout—in the unfriendly soil of Berlin.

TRIUMPH IN BERLIN

Einstein came to Berlin in 1914—shortly before the First World War broke out.

The Germans had made him an offer hard to refuse: high salary, no teaching or other duties, and a prestigious academic position at the newly established Kaiser Wilhelm Society for the Encouragement of Science. Moreover, living in Berlin was his divorced cousin Elsa Einstein, whom he had known since childhood and with whom, his letters to her suggest, he was having an affair. Yet he had hesitated before accepting the position. As a child in Germany he had disliked the disciplined German temperament, with its underlying militarism, and he still did.

Arriving in Berlin with misgivings, he wrote to a friend, "These cool blond people make me feel uneasy; they have no psychological comprehension of others."

He immersed himself in his attempt to generalize the theory of relativity so as to include the force of gravity. Building on Minkowski's ideas, he proposed the bold idea that gravity results from the warping of the four-dimensional fabric of spacetime by massive objects. The theory, he said, could be checked during an eclipse of the sun. If it were true, light from a distant star that passes near the sun in its journey to earth would be deflected toward the sun by the warp in spacetime created by the sun's mass. That could be observed only during an eclipse, which blocked the sun's glare.

Such an eclipse occurred in 1919. Astronomers excitedly found that Einstein was right. Confirmation in the distant heavens of the thinking of an earthbound man was heralded in newspaper headlines on every continent. The First World War had ended. A wounded and divided world hungered for something that could be shared by all. Einstein became an instant celebrity.

Today, many decades later, long familiarity has dimmed our sense of awe at Einstein's achievement. An observation by Woody Allen puts it back in perspective: "Can we actually 'know' the universe?" asked Allen. "My God, it's hard enough finding your way around in Chinatown."

Einstein, who undoubtedly would have gotten lost in Chinatown, had impressed not only the world but also himself. "I want to know," said a now cocky Einstein to one of his student friends,

Esther Salaman, "how God created this world. I'm not interested in this-or-that phenomenon, in the spectrum of this-or-that element. I want to know his thoughts, the rest are details."

Now it is true that Einstein was talking to an attractive young woman whom he may have wanted to shock slightly, but his words do indicate his mental state. "A priori one should expect a chaotic world which cannot be grasped by the mind in any way," he marveled. He had always believed in the unity of the physical world. He now believed it was a unity potentially understandable by man and so providing evidence that the universe was not alien but held human meaning.

Though he had supplanted the warm, comforting God of his childhood with the cold, rational one of science, he had found in this apparently impersonal universe a place for the human mind.

Einstein on the Miracle of the World

Einstein's feeling of awe for the mystery of reality never faltered through the years. When he was
seventy-three, he wrote to his friend since student
days, Maurice Solovine:

. . . You find it strange that I consider the com-
prehensibility of the world (to the extent that we
are authorized to speak of such a comprehensibil-
ity) as a miracle or an eternal mystery. Well, a
priori one should expect a chaotic world which
cannot be grasped by the mind in any way. One
could (yes *one should*) expect the world to be sub-
jected to law only to the extent that we order it
through our intelligence. Ordering of this kind
would be like the alphabetical ordering of the
words of a language. By contrast, the kind of order
created by Newton's theory of gravitation, for instance, is wholly different. Even if the axioms of
the theory are proposed by man, the success of
such a project presupposes a high degree of ordering of the objective world, and this could not be
expected a priori. That is the "miracle" which is
being constantly re-enforced as our knowledge
expands.

There lies the weakness of positivists and pro-
fessional atheists who are elated because they feel
that they have not only successfully rid the world
of gods but "bared the miracles." Oddly enough,
we must be satisfied to acknowledge the "miracle"
without there being any legitimate way for us to
approach it. I am forced to add that just to keep
you from thinking that—weakened by age—I have
fallen prey to the clergy. . . .

3

"Nobody Knows What Time Is Anyway"

Just as people find water whenever they dig, man everywhere finds the incomprehensible sooner or later.
GEORG CHRISTOPH LICHTENBERG

If you do not really believe the theory of relativity describes the world about you—for instance, that time can flow differently for an entity moving at a different speed from your own—do not despair. You are in very good company.

When young Albert Einstein had struggled with the concepts that would lead to the special theory of relativity, he discussed them while sipping drinks in cafés, smoking cigars in his rooms, hiking in the mountains, with his good friend, Michele Besso. Besso tried to understand—he questioned and probed as Einstein explained his ideas.

Einstein finally presented his theory in a deceptively unpretentious paper entitled *On the Electrodynamics of Moving Bodies*. Unlike most scientific papers, it was not studded with references—there were none at all. But in the paper he did acknowledge the help of one person: Michele Besso. Yet in February 1952, after almost a half-century of correspondence, much of it on physics, Einstein wrote to Besso chiding him for not really accepting the theory of relativity in his own understanding of the nature of the world!

Generally speaking, such has been the fate of Einstein's new view of time in the special theory of relativity: people, including those who think deeply, pay it lip service but do not really accept its tenets as part of their inner being.

Most physicists and astronomers don't. They routinely make calculations using relativistic formulas that correctly predict the course of subatomic or astronomical events. Yet, though they have assumed relativity theory's truth at the levels of the very small and the very large, they blithely disregard its relevance in understanding the middle-sized universe that is their home.

Most philosophers don't. Einstein's walking companion during his later years was another lonely thinker of impressive accomplishment, logician Kurt Gödel. Gödel, whose work had demolished much of the existing philosophical foundation of mathematics, expressed surprise at how little relativity theory had affected his philosophical colleagues' understanding of reality. It very much influenced his own.

Albert Einstein did accept the truth of relativity theory as an intrinsic aspect of his thinking. To him its confirmation during the 1919 eclipse confirmed something more: the physical world is one of being, not of flowing, and can be conceptually understood as a separate entity from the observing human mind. Therein lay the nucleus of a situation which some have called tragic. His stubbornness, or greater vision, depending on your viewpoint, led to growing estrangement from work being done by other theoretical physicists—estrangement to the extent that in the latter half of his life Einstein found himself isolated from the mainstream of physics' development. That is a topic for a later chapter.

Why do not the findings of the theory of relativity underlie most educated people's understanding of life? It is because the physical effects predicted by relativity theory can be discerned only when very high speeds are involved—speeds that are an appreciable fraction of the speed of light. Hence, our earth-bound living experience does not lead to it. The concept challenged by relativity—the belief that physical time flows evenly, under all circumstances the same for you and me and for all other beings who may exist in our universe—is a "truth" so deeply burrowed within our thought processes that we remain unaware how completely it shapes the way we think.

We find it hard to resist giving universal significance to the word "now." But if there is no common universal time flow, there can be no common "now." We may speculate on whether an intelligent civilization has ever arisen on a planet rotating about a distant star and then wonder if intelligent beings are now alive on it. But "now" on that distant planet has no correlation with "now" on Earth, so the latter question has no meaning.

To realize the poignancy possible in this situation, consider one of those residing on that faraway planet to be your father who journeyed there some years ago on a spaceship of advanced design, a ship able to make a one-way trip through a byway, which cosmologists call a "wormhole," connecting two otherwise distant points in the curved fabric of spacetime. Is he now alive? Since you are referring to your own "now," the answer to the question cannot be "yes," "no," or even "maybe." The answer can only be that the question has no meaning.

Besso's error, for which Einstein chided him, was his assumption that his "now" represented the world's full materiality. Einstein wrote to him: "You do not take seriously the four-dimensionality of reality but . . . consider the present the only reality." What Besso considered past and future, Einstein continued, "are physically spatial sections to which the theories of relativity give objective reality." Though he could not accept it in the bowels of his being, the flow of Besso's "now" in space and time was unique to himself and not part of the objective world.

Time flows through a "now" for the perceiving individual but not for the physical universe. Einstein commented on the puzzle of human time in a lecture at Princeton:

The experiences of an individual appear to us arranged in a series of events; in this series the single events which we remember appear to be ordered according to the criterion of "earlier" and "later," which cannot be analyzed further. There exists, therefore, for the individual, an I-time, or subjective time. This in itself is not measurable.

We are touching on a mystery that Einstein considered to be impenetrable: the meaning of "I"-ness with its feelings of time-flow and "now"-ness. There is neither flow of time nor "now" in the theory of relativity and he believed them to be outside the physical picture of the world. Philosopher Rudolf Carnap, who discussed these matters with Einstein, reports: "the problem of the Now worried him seriously. He explained that the experience of the Now meant something special for man, something essentially different from the past and the future, but that this important difference does not and cannot occur in physics. That this experience cannot be grasped by science seemed to him a matter of painful but inevitable resignation."

Not only is there no moving "now" in the equations of physics; the equations can be used equally well in either direction: from what our consciousness considers to be from "earlier" to "later" or from what we would think of as "later" to "earlier." Physicist Richard Feynman has made practical use of this phenomenon in making subatomic calculations. He obtains accurate results by considering a particle of matter that has the opposite physical charge of ordinary matter to be ordinary matter moving in reversed time—that is, moving from the particle's "future" to its "past."

So let us not blame Besso—or ourselves—for not fully fathoming time. Einstein is telling us that there are human limits to our ability to understand. He once asked the pioneer child psychologist, Jean Piaget, to investigate how young children developed their sense of time: "Is our intuitive grasp of time primitive or derived?" Einstein—the man who penetrated the physical meaning of time more deeply than any earlier scientist in human history—exclaimed in the last few months of his life, "Nobody knows what time is anyway!"[1]

THINKING IN TERMS OF A WORLD WITHOUT "NOW"

"According to the theory of relativity, the future is as real as the past," said Einstein. He wrote, "For myself there exists in the final analysis no becoming, but only being."

By "being" Einstein means four-dimensional being in spacetime within which three-dimensional change can, and does, take place. For example, Einstein says, "This naturally does not mean a negation of a gradual development of the organic world on the surface of the earth."

---

1. He was responding to a query about the amount of time required to create the world.

But to try to mentally step out of the flow of time and think of the world as pure being, with no moving "now" that identifies a present moment separating past from future, takes practice—and may be impossible to fully do since the focus of that moment in which we are doing our thinking, "now," is always there. Let us consider an example from science fiction involving time travel.

There may be valid theoretical ways to travel into the past—please note that important word *theoretical*—which depend on relativity theory. Einstein's logician friend, Kurt Gödel, proposed one, fifty years ago. More recently, the cosmologist Kip Thorne at Caltech has suggested a way based on passage through a wormhole in spacetime.

In a familiar plot, the hero (sometimes, heroine) goes back in time and proceeds to fall in love and engage in other equally hazardous pursuits. He must be careful, since this can lead to logical complications if he changes the course of history— perhaps to the extent that the hero never gets to be born. Sometimes that is the purpose of the trip back in time: to change the past so that it results in a more pleasant present.

Writers of such fiction often resort to the theory of relativity to make the junkets back in time seem credible. But in doing so, if they permit the past to be changed, they are showing they do not understand what Einstein has tried to tell us: the world is one of "being," not "becoming." They have not succeeded in changing their own views of reality so that it jibes with that of the theory of relativity.

Einstein's "world of being" is not one of "becoming," but it is also not one of inactivity: It is one in which what we perceive to be movement and the passage of time are neither, but are merely different locations within four dimensions of spacetime reality. It is a world of "being" because events in those four dimensions are fixed. And it is not a world of "becoming" because it is not sliced by a universal, moving, knife-edge-like "now" separating past from future. There are no past and future, but only "earlier" and "later," in the four-dimensional spacetime universe. In Einstein's words, "Physical time knows only a time-bound order of the situation but no 'now,' no 'past,' and no 'future.'"

The science fiction hero or wormhole traveler may go back in time and wreak havoc there. But he is already a product of that havoc which may even be in his history books where its source goes unrecognized. The opposite statement is also true: the traveler who goes back in time cannot modify the past while he is there because in the world of spacetime being he has already done so: The past never had the opportunity to be unmodified. The two opposing ways of expressing the matter—only superficially paradoxical—are due to both our thought and language reflecting a non-Einsteinian way of looking at the world.

We are trying to reason about events in a world in which time does not flow using our already-established thought processes, which implicitly assume it does. No wonder we are easily confused.

QUESTIONS OF GOD, MAN, AND UNIVERSE

This different manner of thinking about the world puts a different slant on questions about its origin and fate. Did God create the universe? If not, how was it created? What was happening before then? Will all that humanity accomplishes be lost when the universe decays into a heat-death that can no longer support life?

Einstein wrote:

> The sense of the religious, which is released through the experience of potentially nearing a logical grasp of these deep-lying world relations, is . . . a feeling of awe and reverence for the manifest Reason which appears in reality. It does not lead to the assumption of a divine personality—a person who makes demands of us and takes an interest in our individual being. In this there is no Will, nor Aim, nor an Ought, but only Being.

Just asking the questions, with Einstein's world of "no Will, nor Aim, nor an Ought, but only Being" in mind, suggests the answers. The universe wasn't created; it just is. Scientists who discuss the origin and evolution of the universe do so in terms of an initial event. They do not tell us how that beginning event came into being. We cannot ask what happened before the world was created because the words "create" and "before" refer to time, and time does not exist if there is no world.

Unfortunately, these are not psychically satisfying answers: they do not satisfy our emotional need to seek knowledge of ultimate truths. Can we use the word "creation" in an ultimate sense—one beyond time? Only if we drape its meaning in mystery beyond our ken, because it then becomes beyond our ability to visualize. Interestingly, if we do so we are taking our lead from the Hebrew Bible, which uses a special word for "create" when it refers to God's creation of the world, a word whose meaning we cannot compare or explore further since the Bible never uses it in any other context. If the word "create" is beyond our ability to comprehend, the concept "God the creator" becomes an even deeper, even more impenetrable mystery.

Will whatever levels humanity manages to reach be lost when the universe can no longer support life? Those who ask this question tend to respond "yes," and then add that it is a somewhat depressing thought. Actually, it is a rather optimistic thought since it assumes that mankind will last that long without blowing itself up. But again, if we look upon the world from a standpoint of "being" rather than "becoming," human history is not lost because nothing is lost: all continues to be. (Even the word "continues" suggests movement in time and in so doing indicates our predicament: we cannot get away from our time-flow-based mental processes.)

But there is an even more remarkable implication of the world of being, without a universal "now," that most people who fully accept the theory of relativity tend to block from their minds. Consider the case of an astronaut who is returning to Earth after an extensive trip in a rocket traveling at a rate that is an appreciable percentage of the speed of light. Masanao Toda, of Japan's Hokkaido University, addresses the situation:

> The astronaut lives the local time of his spaceship, and his travel will be represented by his now-point constantly moving on the time-line characteristic to the ship, while the now-point of the system Earth will proceed on its time-line according to the local rate. This is as if two boats leaving the port together are going into two different channels. Is there any guarantee that the boats will meet exactly at the point where the channels join again? How can one be sure that, when the astronaut's now-point reached the temporal moment when the spaceship and Earth are scheduled to meet, the now-point of Earth, too, will exactly reach the corresponding temporal spot?

One can't. The great majority of educated people today accept both the probability of higher-speed space travel in the future and the truth of relativity. But in doing so they thoughtlessly assume the four-dimensional spacetime universe somehow has a single, alive now-point, specifically the "now" in which they live, that represents "true" reality as it moves into the future. But if that were the case, Toda reminds us, the astronaut might meet the earth "where events have not yet happened, or the earth where all the events have already happened!"

In considering this situation of a spaceship returning to Earth and to "now" from a flight near the speed of light, it becomes clear that *every* instant on Earth since it left must be alive—ready to be "now" to the returning astronauts.

Perhaps a spaceship took off from Earth in prehistoric times and has not yet returned. (It may have been the product of an advanced civilization since destroyed by nuclear catastrophe.) The same reasoning applies. To accept the theory of relativity is to believe that every moment in the past and future of conscious man is alive.

This realization is either exhilarating or depressing depending on your viewpoint and the nature of your life experiences. Its implications for understanding Einstein's thought—and for understanding the world itself—are profound.

4

Heritage Regained

Perhaps in time the so-called Dark Ages will be thought of as including our own.

GEORG CHRISTOPH LICHTENBERG

Einstein took a sardonic view of his unexpected world popularity after the eclipse in 1919 confirmed his theory. Lecturing in Paris at the Sorbonne, he said, "If my theory of relativity is proven successful, Germany will claim me as a German and France will declare that I am a citizen of the world. Should my theory prove untrue, France will say that I am a German, and Germany will declare that I am a Jew."

He enjoyed humorous stories and was particularly fond of Jewish jokes, preferably dirty ones.[1]

---

1. Unfortunately, I have not been able to locate any specific examples.

Like other great comedians, he was not above using the same joke—with a twist angled toward the audience he was addressing—more than once. His friend, John Plesch, reports that he himself would considerately laugh a second time at a joke he had heard before. Some time after speaking in Paris, he wrote in a letter to the London *Times*: "Today I am described in Germany as a 'German savant,' and in England as a 'Swiss Jew.' Should it ever be my fate to be represented as a *bête noir*, I should, on the contrary, become a 'Swiss Jew' for the Germans and a 'German savant' for the English."

GERMAN ANTI-SEMITISM

But, as Einstein had become grimly aware, anti-Semitism was no joking matter in a defeated, post-war Germany. He wrote to Paul Ehrenfest, who had found a position in Holland at the University of Leyden, "Here strong anti-Semitism and mad-dog reaction reigns—at least among the so-called cultured."

Among the "so-called cultured" were two fellow Nobel Prize–winners in physics: Philipp von Lenard (who may have been of Jewish ancestry) and Johannes Stark, both viciously anti-Semitic. The two distinguished professors denounced Einstein's ideas about relativity as "Jewish physics." Lenard was at Heidelberg University, where he even refused to allow Jewish students to work in his physics laboratory.

Nevertheless German Jews tended to be intensely patriotic, proud of their long history in the country. Since their emancipation from the ghetto, they had become prominent beyond their numbers in the business world and in cultural life.

The situation made Einstein uneasy. He remembered that in Prague, too, Jews tended to dominate the city's cultural affairs. An intellectual, he thought it would be better if fewer Jews were intellectuals. At Heidelberg, for example, a friend of Einstein's wrote, "There were so many baptized and nonbaptized Jewish professors that the anti-Semites used to say, 'One-half the professors are Jews and the other half baptized Jews.'" In 1929 Martin Heidegger, well known from the publication two years earlier of his treatise *Being and Time*, wrote to the education ministry warning them against the "growing Judaisation" of the universities.

Making the situation even worse, more Jews were pouring into Berlin from the small Jewish villages and crowded ghettos of eastern Europe. Of foreign dress and foreign behavior, they had an intense desire to better themselves that came out as aggressiveness. They embarrassed the insecure German Jews, who were trying to act in a manner indistinguishable from other Germans. Though Einstein was uncomfortable with the pervasive Jewish presence in Berlin's cultural life, he was more troubled by the self-hatred that made German Jews treat the immigrants so inhospitably.

He himself saw good qualities in the unculti-
vated east European Jews. He approved of their
ambition, turn of mind, pride in being Jewish. He
saw the common nature of his own intellectual
striving and that of the new immigrants. He said
that if he had been born in eastern Europe he would
probably have become a rabbi.

A TROUBLED LIFE IN BERLIN

That comment is less strange than it might appear.
Events in his own life had made him realize that
science alone cannot solve the existential problems
of human living. With the certainty of youth, he
had discarded the religious worldview of his childhood and replaced it with a scientific one. But his
years of greatest achievement—the years during
which he was completing his general theory of
relativity and impatiently anticipating the eclipse
that he was certain would confirm its truth—had
been years of war and personal trial.

He had come to Berlin with his family on the
eve of the First World War. Shortly thereafter, his
wife, Mileva, terminating a stormy relationship,
moved back to Zurich with their two sons—Hans
Albert and Eduard—whom he loved. His younger
son, Eduard, had been diagnosed as schizophrenic.
The war had added to his distress. Horrified by the
mass killing, he opposed the fighting and was put
in the painful position of hoping that the country
in which he lived would be defeated. He wrote to
his friend Paul Ehrenfest:

I cannot help being constantly very depressed over the immeasurably sad things which burden our lives. It no longer even helps, as it used to, to escape into one's work in physics.

He became more aware of himself as having human feelings and a human conscience—a realm not well served by science. As he put it years later:

Most scientists treat conscience as a stepchild in their picture of the world. This is a kind of disease of the profession and one should always be conscious of this weakness.

Now, in the postwar years, his greatest work accomplished, he became less single-mindedly devoted to science. He and Mileva were officially divorced in 1919, freeing him to marry his cousin Elsa after their long affair. Elsa, divorced since 1908 from her first husband, came to Einstein with two young daughters, Eva and Margot, giving him a family once again.

THE ZIONIST CALL

In 1919 also, a leader of the Zionist movement in Germany, Kurt Blumenfeld, called on Einstein to seek his endorsement. Blumenfeld later reported the questions Einstein asked him, questions that reveal the ideas Einstein then held about being Jewish. Here are the questions:

. . . Is it a good idea to eliminate the Jews from the spiritual calling to which they were born? Is it not a backward step to put manual capabilities, and above all, agriculture, at the center of everything Zionism does? . . .

Are not the Jews, through a religious tradition which has evolved outside Palestine, too much estranged from the country and country life? Are not the talents which they have exploited with such scientific accomplishment perhaps the result of an innate spirituality? . . .

Einstein seemed remarkably preoccupied with Jewish spirituality in the questions he asked.

Blumenfeld did eventually convince him. "Zionism," Einstein wrote to Paul Ehrenfest, "represents a new Jewish ideal that can restore to the Jewish people their joy in existence."

He had now returned to the Jewish fold—but not wholly. Under German law the people were divided into religious groups, each of which legally taxed its members. When the organized Berlin Jewish community billed him for membership in 1920, Einstein refused to pay. "Much as I feel myself a Jew," he said, "I feel far removed from the traditional religious forms." He offered instead to contribute each year to the Jewish community's welfare program. "Nobody can be compelled to join a religious community," Einstein proclaimed. "Those times, thank God, are gone forever. I hereby declare once and for all that I do not intend to join. . . ." Three years later he joined.

In Zionism, Einstein found reconciliation with his Jewish background. As a Jewish intellectual he had earlier felt himself alienated from the body of the Jewish people. In 1922, he put it this way:

> What is happening today among the Jewish nation, a short time ago no one would have thought possible. While up to now the intellectual elite of the Jews turned away consciously or unconsciously from the larger group of Jews due to our lack of resonance, now our nation suddenly stands in front of us again with a freshness of life and each of us loses the feeling of loneliness.

He visited Palestine in 1923 and saw what Zionism had accomplished. There were some eighty-five thousand Jews living there then. He was invited to deliver the very first lecture at the Hebrew University on Mount Scopus. Unfortunately there was no Hebrew University on Mount Scopus as yet—the building of the campus would not begin until two years later. But that little matter troubled nobody. They did have a house.

Orated the official who introduced him: "Three thousand years ago, King Solomon built a house to the Lord of the world on Mount Moriah, and his first prayer in this house was that this should become a house of prayer for all peoples. Now, as we are building this house, we pray that it should become a house of science for the whole world. Professor Einstein, please step up on the stage which has been waiting for you these past two thousand years!"

Einstein, in accordance with the historic nature of the occasion, began his lecture in careful Hebrew:

I, too, am happy to read my address in the country from which the Torah and its light emanated
to all the enlightened world, and in this house
which is ready to become a center of wisdom and
science for all the peoples of the east. I regret being
unable to deliver my speech in the language of
my people, and I must continue in the language
understood by all of you, the French language.

He then talked about relativity. He seems to
have had the type of childhood Hebrew education
common in America today, judging by these words
that he later wrote in his travel diary: "I had to
begin with a greeting in Hebrew, which I read with
great difficulty."

Emotionally moved by what he saw in Palestine, he spoke of his feelings to a cheering crowd
at *Lemel*, a children's school:

I consider this the greatest day of my life. Hith-
erto I have always found something to regret in
the Jewish soul, and that is the forgetfulness of its
own people—forgetfulness of its being almost.
Today I have been made happy by the sight of the
Jewish people learning to recognize themselves
and to make themselves recognized as a force in
the world. This is a great age, the age of the lib-
eration of the Jewish soul.

RETURN TO THE JEWISH PEOPLE

He returned to Germany now deeply and willingly
involved in Jewish affairs. He helped raise money

for the Berlin Jewish community. He joined the Board of Governors of the Hebrew University, writing to Paul Ehrenfest, "I do believe that in time this endeavor will grow into something splendid, and, Jewish saint that I am, my heart rejoices."

Most comfortable in Jewish company, he once told a prominent journalist that he knew the reporter was Jewish because he felt at ease being interviewed by him. As it happens, the man was not. When told this, Einstein retorted that he must have Jewish ancestors. His closest friends, both male and female, were Jewish. To one of them, young Esther Salaman, he made a significant remark:

> I don't love the Germans but my reasons are probably different from yours. . . . I don't know how to express it. . . . I've found the answer. They're not religious.

Being a fair-minded equal-opportunity critic, he had this to say about the spirituality of his fellow German Jews in 1929:

> The greatest enemies of Jewish national consciousness and Jewish dignity are fatty degeneration—by which I mean the loss of moral fiber which results from wealth and comfort—and a kind of spiritual dependence on the surrounding non-Jewish world which is a consequence of the disruption of Jewish corporate life.

By 1930 the handwriting was on the wall in Germany—and it was in the form of swastikas. Eu-

ropean Jews found their situation rapidly dete-
riorating. In that year, Einstein spoke to an assem-
bly of British Jews words that could have come
from a biblical prophet:

> To you I say that the being and fate of our people
> depends less upon external factors than that we
> remain true to our moral traditions which have
> carried us through the centuries in spite of the
> heavy strains which broke in upon us. In the ser-
> vice of Life sacrifice becomes a grace. Within the
> traditions of the Jewish people exists a striving
> towards righteousness and understanding that
> should be of service to the rest of the nations both
> now and in the future.
>
> Do not bemoan the hardness of fate, but in this
> occurrence see rather a motive for both being and
> remaining faithful to the Jewish community.

"In the service of Life," said Einstein, "sacrifice becomes a grace." That can be the statement
only of a deeply religious man.

A JEW IN PRINCETON

In early 1933, when Hitler came to power, Einstein
and his wife, Elsa, were on a trip to the United
States. He never returned to Germany but settled
in Princeton, New Jersey—the first professor at the
new Institute for Advanced Study. There he would
stay for the rest of his life.

He was now fifty-four years old, and his greatest work had been completed many years earlier.

Though he was now laboring on a generalized theory that would unite the electromagnetic and gravitational fields, the phenomenal instinct for the correct paths to follow now eluded him. Unable to repeat the successes of the past, he devoted more time to nonscientific pursuit, including endeavor for the Jewish people.

"I must be a saint for the Jews, and bestow my blessing on all the *goyim*," he wrote from Princeton to a friend in Israel. He took some pride in that status, often jokingly but semiseriously referring to himself as a "saint for the Jews." He did not—quite rightly—feel like a saint. The self-mocking tone is typical. Einstein laughed much, often at himself.

The Institute for Advanced Study was the idea of a prominent educator, Abraham Flexner, who named himself as its head. Flexner had gained deserved renown for producing an acute and much needed critique of American medical training and then obtaining the money from wealthy philanthropists to reform it. He obtained funding for the new institute from Mrs. Felix Fuld and her brother, Louis Bamberger, former owners of Bamberger's department store in Newark. Wealthy from having sold their store to Macy's, and wanting to support medical education, they approached Flexner to tell them how to do it. But instead he persuaded them to establish a place for distinguished scholars to study and work free from the need to teach pesky students. His institute soon attracted noted mathematicians and scientists.

Many of them—as were the Bambergers, Flex-

ner, and Einstein—were Jewish. *Time* magazine
editor T. S. Mathews, whose home at the time was
near Einstein's, writes: "The small, complacent,
parochial community of Princeton began to wrin-
kle its anti-Semitic nose. When the institute bought
a tract of land on the outskirts of the town and put
up some modest buildings, this area was promptly
dubbed 'the new Jerusalem.'"

In Berlin, in 1921, Einstein had said about his
Jewish colleagues: "I have always been annoyed
by the undignified assimilationist cravings and
strivings which I have observed in so many of my
friends." Now he was dismayed to observe similar tendencies in almost his own backyard.

Abraham Flexner, the man who had hired him,
had managed to write a long autobiography without ever mentioning the words "Jew" or "Jewish."
In it he refers to his parents as "pious" without
divulging the religion in which they were pious.
Flexner warned Einstein that Jews must be careful not to cause the rise of anti-Semitism in the
United States. But the warning clearly did no good.
Flexner later told a Newark rabbi that Einstein's
behavior was such that it would be impossible for
him to invite any more Jews to Princeton.

One of Einstein's "crimes" was wanting to
give a violin recital for the benefit of Jewish refugees. Flexner said Einstein was being "driven by
pure vanity" in doing such things, adding, "I know
him well." He once wrote Einstein an appeasing
letter in which he referred to himself as a "native
American" and in which he quoted Voltaire. Ein-

stein, clearly enjoying tweaking Flexner's nose, responded with a wry note:

> Your letter from the 12th gave me special joy, #1, because it came from Abraham Flexner; #2, because it was written by a kosher American; #3, because you called upon the not-so-kosher Voltaire as your witness.
> With much thanks and greetings.

Voltaire, it should be remembered, is the man who in his *Dictionnaire Philosophique* called Jews "the most abominable people on Earth."

When his old friend Michele Besso wrote from Europe that he was converting from Judaism to Christianity, Einstein reproved him with a touch of humor: "You are a good man," Einstein wrote back, "so you will certainly not go to hell, even though you have asked to be baptized." But, he said to Besso, he did see one advantage in the conversion: "as a *goy*, you need not feel compelled to learn Hebrew." Einstein said that he, himself, did feel he should learn Hebrew, but was resisting doing so and feeling guilty about it.

RETURN TO JEWISH TRADITION

Einstein judged Germany's Jews harshly. He revealed his feelings privately in a letter he wrote in 1935, before the full horror of the Nazi effort was known:

There is no doubt that the misfortune of the German Jews springs in large part from the fact that they, to a large degree, became unfaithful to the Jewish community and tradition.

He himself had wholeheartedly returned to that tradition. He wrote some months later:

I believe that it would be a real loss if the Jewish tradition as such went out of being. I find that the ideal of an openhearted humanness is better incorporated in our Jewish tradition than in the tradition of any other such community which brings together so many individuals. I believe too that by keeping our Jewish tradition, it serves the ideal of humanity.

His severe judgment of German Jews as being responsible for their own misfortune because they had become "unfaithful to the Jewish community and tradition" was cruelly biblical. His statements that the Jewish people "serves the ideal of humanity" and "contribute to the ennoblement of the human race" hark back to the biblical covenant God made with Abraham.

He believed that the emotion inspiring his own scientific quest to be the same as that voiced in biblical psalms. It is, he writes, one of "joy and amazement at the beauty and grandeur of this world, of which man can form just a faint notion. This joy is the feeling from which true scientific research draws its spiritual sustenance."

He had returned to the Jewish people and to Jewish tradition. But had he returned to the Jewish God?

Einstein's "Wicked Tongue" on Intolerant Clergymen

"My sweetheart has a very wicked tongue," Mileva confided to a friend before she and Einstein were married. He never lost it. He wrote to Rabbi Solomon Goldman from Princeton:

> . . . A man who is convinced of the truth of his religion is indeed never tolerant, and he is unable to be tolerant. At the least, he is to feel pity for the adherent of another religion but usually it does not stop there. The faithful adherent of a religion will try first of all to convince those that believe in another religion and usually he goes on to hatred if he is not successful. However, hatred then leads to persecution when the might of the majority is behind it.
>
> In the case of a Christian clergyman, the tragic-comical is found in this: that the Christian religion demands love from the faithful, even love for the enemy. This demand, because it is indeed super-human, he is unable to fulfill. Thus intolerance and hatred ring through the oily words of the clergyman. The love, which on the Christian side is the basis for the conciliatory attempt towards Judaism is the same as the love of a child for cake. That means that it contains the hope that the object of the love will be eaten up. . . .

When the Union of Orthodox Rabbis expelled Rabbi Mordecai Kaplan because he expressed disbelief in a personal God, Einstein said of them:

"To take those fools in clerical garb seriously is to show them too much honor."

5

"We Are Like a Child Who Judges a Poem by the Rhyme"

Man is a masterpiece of creation, if only because no amount of determinism can prevent him from believing that he is a free being.

GEORG CHRISTOPH LICHTENBERG

Is man a free being? (Of woman I dare not speak.) Or are his thoughts, feelings, and decisions at any moment determined by physical events in the world—including those that have resulted in the structure of his brain at that moment? The choice as to which is true has inspired heated dispute over the centuries. We instinctively feel that if we knew the answer it would help us understand ourselves. How did Einstein answer? He gracefully sidestepped by simply answering yes to both questions.

He firmly believed in a physically determined universe—that is, the four-dimensional world of being consists of an interrelated whole in which each happening is caused by the physical events relating to it. "Our actions," he wrote, "should be based on the ever-present awareness that human beings in their thinking, feeling, and acting, are not free but are just as causally bound as the stars in their motions."

But in writing to Michele Besso about his younger son's mental illness he exclaimed: "I am to blame for him and I reproach myself!" In the fully determined world in which he believed how can anyone, including himself, be held to blame for anything?

In discussing the matter he liked to refer to Baruch Spinoza. Three centuries ago, at the dawn of the scientific age, that reclusive Dutch Jewish philosopher expressed belief in a completely deterministic world and explored its consequences. Einstein wrote:

> From the point of view of nature [Spinoza would say God] there is no freedom of will. From the point of view of man there exists the illusion of freedom of choice. . . . The pressure of responsible behavior makes this illusion of freedom compatible with causality. Essential factors like tradition and education have to be included in the causality chain.

Nevertheless Einstein believed that a person is responsible for what he does. He argued in a letter:

. . . you make the statement that a consequent deterministic opinion is really antiethical and that therefore an ethical attitude is dependent on the theory of the freedom of the will. That means that it presupposes an incompleteness of the causal connection. . . . Indeed I believe that we should not make the fight for our ethical beliefs dependent on this scientific subtlety. . . .

Consequently his statements concerning answerability for one's actions were blatantly contradictory. Consider the two below:

In 1946 [when Einstein was asked to rejoin the Bavarian Academy, from which he had been made to resign in 1933]: "The Germans have slaughtered my Jewish brethren. I will have nothing further to do with them, not even with a relatively harmless academy."

In 1948: "For me, however, the intellectual base is faith in unlimited causality. I am not able to hate him because he ought to do what he does! I am closer to Spinoza than to the prophets. It is why there exists no sin for me."

He could even express contradictory feelings in the same statement, and he knew it. In a 1944 letter he wrote:

Nevertheless I find it nice and interesting to live and work, in particular because the Germans are being heavily bombed. Why not say so frankly. Formerly I would never have believed that a feeling of revenge could fill me to that extent. I also know that it is stupid because I know the [here Einstein used a crude derogatory epithet] are the way they are because God unfortunately made them that way—provided he ever personally took part in this business.

Einstein's response is, of course, heavily emotional. But still, how can a brilliant scientist, noted for the clarity and logical depth of his thought, apparently contradict himself again and again in the manner of these quotations?

First, bear in mind that free will is semantically not the opposite of determinism; random occurrence, or indeterminism, is its opposite. Einstein believed that a mental event, a thought, corresponds to a physical change in the brain. If there is a break in that causal chain within the brain so that an event is uncontrolled, it represents the introduction not of free will but of randomness. What then is "free will"? The term now lacks real meaning, and Einstein tended to lose patience with those who used it:

Honestly, I cannot understand what people mean when they talk about the freedom of the human will. I have a feeling, for instance, that I will something or other; but what relation this has with freedom I cannot understand at all. I feel that I will to light my pipe and I do it; but how can I connect this up with the idea of freedom? What is behind the act of *willing* to light the pipe? Another act of willing? Schopenhauer once said: *Der Mensch kann was er will; er kann aber nicht wollen was er will.*

His quotation from Schopenhauer translates into "Man can do what he wills but he cannot will what he wills." Nevertheless, Einstein believed that a person is responsible for what he does in a frozen world of four-dimensional being. In one of his many comments on Spinoza in his letters, he wrote:

> If I understood you correctly, you are worried by the conflict between the purely causal attitude of Spinoza and the attitude which is directed to active effort in the service of social justice. I don't think that there exists a real conflict because our spiritual tensions, not only those of the passions but also those of the urges to achieve a more just order of society, belong to the factors which, with all others, partake of causality. It doesn't represent an inconsistency if we connect those spiritual states with the idea of purpose and goal. The conviction that human existence is temporally limited in contradistinction to general cosmic appearances doesn't change anything in the fact that it lies in our nature to act in the direction of an aim.

But Einstein in this letter seems to be contradicting his comment quoted earlier approving Schopenhauer's stand. Is he?

WHY EINSTEIN IS NOT CONTRADICTING HIMSELF

In Einstein's universe a person does not have free will, but his responsibility remains. Physical determinism in the theory of relativity is easily misunderstood: It does *not* imply fatalism. It does *not* imply that the individual has no effect on the passage of events. In a world of being, consisting of a complex network of spacetime interrelations, the creative individual can be looked upon as an essential node which has helped determine the pattern of interrelation. For the individual in such a world, determinism has lost its lethal punch.

This may sound confusing because it is. Let us see why. In an earlier chapter we spoke of the difficulty of purging the idea of the flow of time from our thought processes in order to think in terms of a four-dimensional spacetime universe in which time does not flow. Einstein could have written to many of us what he wrote to Michele Besso: "You do not take seriously the four-dimensionality of reality but . . . consider the present the only reality." *It is the unthinking combining of two contradictory mental positions that cannot be combined—that of four-dimensional reality and that of time flow in which only the 'now' is real—which leads to the misunderstanding of determinism.*

Picture the four-dimensional spacetime universe that Einstein proposes. This universe is four-dimensionally static—nothing in the picture moves. That the universe may be expanding does not change the static nature of the picture; the expansion is manifested by the portion of the picture representing "earlier" being smaller than the portion of the picture representing "later."

With that picture firmly in mind, notice that causality in such a universe can not mean that the future is fully determined by happenings along a line representing the present moment slicing through it—it can not because there is no such line! The concept that causality consists only of the de-termination of the future by the present is a con-cept left over from a view of reality that sees time as a moving "now." Listen to Einstein:

> We must distinguish between causality as a pos-tulate directed towards theory and causality as a postulate directed towards the observable. This last demand remains unsatisfied—empirical cau-sality does not exist—and it well may remain so. I believe it is too narrow a formulation to consider causality as necessarily a temporal sequence be-tween the present and the future. That is only one form of the causal law—but not the only one. Ac-cording to the general theory of relativity, time loses its independence and becomes only a single co-ordinate of a four-dimensional system called the *world*. In the four-dimensional world causal-ity will be only a link between two breaks. This constitutes causality as it corresponds to the gen-eral theory of relativity.

We cannot expect to comprehend in more than a primitive way the causal pattern of which we each form part. Says Einstein: "We are like a child who judges a poem by the rhyme and knows noth-ing of the rhythmic pattern." Since each part of the great pattern of being in spacetime is dependent on the parts to which it relates, the part under con-sideration itself, if it is not repressed, plays more than a passive role in the determination of the nature of its own being: it helps determine the other parts which in turn determine itself. To illustrate this situation see the simple example, *A Miniworld of Weights on a Balance.* Says Albert Einstein: "External compulsion can, to a certain extent, reduce but never cancel the responsibility of the individual."

To summarize: Einstein's world is a static interrelated set of events in four dimensions of spacetime. There is nothing in it that corresponds to a moving present moment—that concept is a vestige of the time-flow preconception we are trying to remove from our minds. Time flows for the perceiving individual but not for the universe.

If you exist in such a world of being, your striving is part of that pattern of being—a pattern in which the surrounding world affects you, but also in which you, if you are physically and politically free to do so, affect the surrounding world.

Notice that the idea of a deterministic spacetime world can be more exhilarating than depressing. It not only offers room for individual creativity, but offers more: Since it is a world of all-moments-being, rather than of each-instant-fleeting, a world in which the acts of daily living are not ephemeral but enduring, the opportunity to make life meaningful is enhanced. Every act performed is a contribution to the structure of the world—a sobering and humbling thought. Einstein writes:

Spinoza was the first who really applied the thought of deterministic connection of all events consequently to human thought, feeling and ac-

A Miniworld of Weights on a Balance

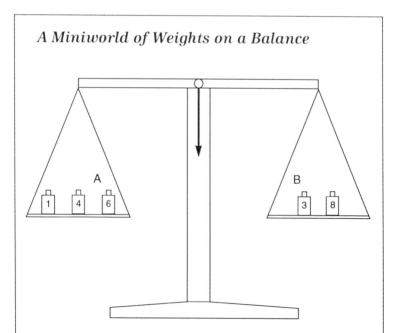

The picture shows a laboratory balance such as is used in a school science class.

The three weights on one side have the same total value as the two weights on the other side, so the balance's pointer stays in the middle and does not move. The balanced balance is a greatly simplified static "world" in which the five weights represent its individuals.

Each individual weight has two active attributes: its mass and which pan it is in. Both attributes are precisely determined by the values and locations of the other weights on the balance in order that equilibrium be maintained—hence it is a deterministic world. Weight A's mass and location, for example, are determined by the masses and locations of the other four weights—including weight B. But weight B's mass and location are also determined by the masses and locations of the other four weights—including weight A.

So in this little deterministic world, each weight partially determines itself.

tion. In my opinion his point of view did not succeed among those who fight for clarity and for what is logically correct because this demands not only consistency in thinking but also extraordinary purity, a greatness of the soul, and humbleness.

EINSTEIN DEFENDS HIS POSITION

Three hundred years after Spinoza, Einstein has been no more successful in convincing others. In a recorded discussion in 1932, he once explained his view on determinism from a historical standpoint. He said that the simple causality which most people assume to be true without giving the matter thought has the same relationship to true causality that "Chopsticks" played on the piano has to a Bach fugue:

> I think that much of the misunderstanding encountered in all this question of causation is due to the rather rudimentary formulation of the causal principle which has been in vogue until now. When Aristotle and the scholastics defined what they meant by a cause, the idea of objective experiment in the scientific sense had not yet arisen. Therefore they were content with defining the metaphysical concept of cause. And the same is true of Kant. Newton himself seems to have realized that this prescientific formulation of the causal principle would prove insufficient for modern physics. And Newton was content to describe the regular order in which events hap-

pen in nature and to construct his synthesis on the basis of mathematical laws.

Now I believe that events in nature are controlled by a much stricter and more closely binding law than we suspect today when we speak of one event being the *cause* of another. Our concept here is confined to one happening within one time-section. It is dissected from the whole process. Our present way of applying the causal principle is quite superficial. We are like a child who judges a poem by the rhyme and knows nothing of the rhythmic pattern. Or we are like a juvenile learner at the piano, just relating one note to that which immediately precedes or follows. To an extent this may be very well when one is dealing with very simple and primitive compositions; but it will not do for the interpretation of a Bach fugue.

He then went on to say:

Quantum physics has presented us with very complex processes and to meet them we must further enlarge and refine our concept of causality.

But, apparently, most quantum physicists were not logical enough, pure enough, great enough, or humble enough. For, as we shall see, to Einstein's dismay they rejected causality altogether.

6

No Closer to the Secret of the Old One

A rather impudent philosopher, I believe it was Hamlet Prince of Denmark, once said that there are more things in heaven and earth than are to be found in our compendia. If the simple fellow, who (as is well known) was not in his right mind, intended to sneer at our compendia of physics, then we can confidently reply to him: Very well but there are also a great many things in our compendia of which nothing is to be found either in heaven or on earth.

GEORG CHRISTOPH LICHTENBERG

"To the Jews I am a saint, to the Americans I am an exhibition piece, and to my colleagues I am a charlatan." In the latter part of that wry self-evaluation Einstein was referring to quantum theory, a field in which he played a lonely role in the world of theoretical physicists. Fortunately it was a role—that of rebel—that he most enjoyed.

"It is a good thing I have so many diversions for otherwise the quantum theory would have sent me into a lunatic asylum," he grumbled in 1921. And thirty years later he concluded that the theory itself belongs there: "This theory reminds me a little of the system of delusions of an exceedingly intelligent paranoic, concocted of incoherent elements of thought."

The quantum theory describes physical events at the subatomic level, a physical world much smaller than that to which human experience accustoms us. As we should expect then—especially after our exposure to relativity theory's dealing with physical values much greater than those to which human experience accustoms us—matter in the subatomic domain acts in ways to which we are not accustomed. It is called quantum theory because subatomic events seem to involve relationships between irreducible unit-quantities of energy that have been given the Latin name *quanta*.

Though Einstein had done much of the early conceptual work in quantum theory, he had become unhappy with its development—it had none of the clear, pure beauty of his own theory of relativity. It declared not only that subatomic matter acts in an eccentric manner in which chance plays a part, but also that it changes its behavior when observed.

OBSERVED WORLD AND REAL WORLD

Einstein believed there is a fundamental reality—an "external real world"—that exists independently of its scrutiny. It is a real world we cannot know directly but only through interpretation and conceptual construction from observation. "Reality as far as it is known to us is and must always be a man-made construct," he wrote.

To Einstein even spacetime was a created reality, real in the sense that it represents in interpreted form a physical actuality which itself remains inaccessible. In the last year of his life he wrote to the physicist David Bohm that he would be willing to replace spacetime in his theories with a different fundamental element, but, he said despairingly, "with what I have no idea."

Our knowledge of material reality is limited by our ability to physically sense and to understand, but liberated in scope by our ability to create.

This corresponds to what we now know about the eye and brain: What we believe we see is the result of organization and interpretation by the brain of certain wavelengths of light from the outside world impinging on our eyes. A person who gains eyesight in adulthood, after being blind from birth, does not at first see normally because his brain, in creating an internal image, has not yet learned to properly interpret the raw information it receives from the eyes.

Quantum theory denies this sharp distinction between observation and outside world, mediated by interpretation. In Einstein's mind quantum theory hopelessly mixed up observation of an event with the event itself—truly hopeless because it insists there is no hope of extricating one from the other. To his chagrin, quantum theorists, in-

Einstein on the Existence of the
Unobserved World—and Nonexistence
of the Observed World

I believe that the external real world constitutes
a foundation which we could not give up. . . . However I believe we need a conceptual world to convert our sensations into something utilizable for
thought. It is an illusion to think that we perceive
the world. When we say that we perceive the
world we have already transformed our sensations—into something conceptual. What our senses
give us can only become a conception of the world
through a conceptual construction. One cannot
affirm therefore that no world exists behind the
observable world, for this observable world itself
does not exist—that is, the world is not given to
us by our senses.

stead of being humbled by this situation as one
would expect, brightly extrapolated from it to declare that they have discovered a new truth: the
external world does not exist separately from its
observation!

The physicist Abraham Pais writes that Einstein once stopped during an evening stroll with
him and pointed to the moon. "Do you really believe," asked Einstein, "that the moon does not
exist when you are not looking at it?"

Einstein's attitude is often talked about pa-
tronizingly by those who write about the subject.
It is said that he wanted to preserve classical phys-

ics, that he became, as he grew older, too set in his thinking to accept the radical conceptual change engendered by quantum theory. It is said that he stayed engaged in a futile search for hidden variables that would turn quantum theory from a probabilistic into a deterministic science.

But those who see Einstein as a scientific conservative refusing to accept conceptual advance do him injustice. Consider that matter of hidden variables. David Bohm, then a young physicist with whom Einstein corresponded, wrote some papers suggesting that there might exist hidden physical characteristics that determine what occurs in the subatomic realm. But Einstein rejected this type of solution, exclaiming, "It looks too cheap to me!" Einstein's approach was more subtle and truly revolutionary: He did not want to patch quantum theory; he wanted to replace it.

Heisenberg's uncertainty principle, for example, is a basic tenet of quantum mechanics. It states that the precise location and momentum of an electron cannot be measured—the more exactly we measure either location or momentum, the more vague in value becomes the other. The reason is commonly given that this is a measurement problem occurring because of the effect of the observer: in making the measurement the observer interferes with the activity of the electron.

But Einstein explains that the answer is not that simple: the indeterminacy is not an indeterminacy of reality, but one of our *image* of reality; it exists because we have reached the limits of applicability of the quantum-theory picture of what occurs.

"Heisenberg's observation does not prove anything against the causality of being," says Einstein, "but only makes an observation regarding the missing completeness of the relationships between empirically determined constants."

The problem is not merely a mathematical one, as some physicists want to believe, but a conceptual one. Einstein told Pauli that though "the description of physical systems by quantum mechanics is incomplete, there would be no point in completing it, as the complete description would not agree with the laws of nature." Quantum mechanics has been successful in that it establishes sets of mathematical rules that can predict events in the subatomic realm. As such it provides eminently useful techniques. But that something can be mathematicized does not mean that its theoretical basis is true. In Einstein's graphic language: "Mathematics is the only perfect method to lead oneself by the nose."

It was Einstein, himself, in the theory of special relativity, who introduced the idea into science that the world can look different to different observers. But in the same theory he constructed an encompassing unity which transcended the views of all observers, a unity that was fully objective. He believed that a fuller understanding of reality than present-day quantum theory represented would also show the independent reality of the objective world at the subatomic level.

Such a new fundamental theory would supersede quantum mechanics in the same manner his theory of gravitation superseded Newton's. Today's quantum theory would then be seen, as Newton's theory is now seen, to be an empirical theory that provides useful answers but does not lead to any deeper understanding of reality. Quantum theory, Einstein declared to Max Born, brings us "no closer to the secret of the Old One."

OF SUCCESS AND FAILURE

Einstein was painfully aware that his contemporaries in physics whom he respected, and who had themselves won the Nobel Prize, believed he had taken the wrong path in rejecting quantum mechanics.[1] He wrote to de Broglie in Paris solely to explain why he had done so. And he wrote Max Born:

> You believe in the God who plays dice, and I in complete law and order, in a world which objectively exists, and which I, in a wildly speculative way, am trying to capture. I firmly *believe*, but I hope that someone will discover a more realistic way, or rather a more tangible basis than it has been my lot to find. Even the great initial success of the quantum theory does not make me believe in the fundamental dice game, although I am well aware that our younger colleagues interpret this

---

1. Einstein's friend and colleague, Kurt Gödel, made the curious remark in a letter to his mother that Einstein's view "is rejected as a wrong way by the vast majority of the physicists, including the Jewish."

as a consequence of senility. No doubt the day will come when we will see whose instinctive attitude was the correct one.

Einstein believed that a shared objective world was necessary for human beings to truly communicate with each other. Harvard professor Gerald Holton found a fragment in the archives at Princeton in which Einstein expressed his instinctive discomfort with quantum theory:

> It is the postulation of a "real world" which so-to-speak liberates the "world" from the thinking and experiencing subject. The extreme positivists think that they can do without it; this seems to me to be an illusion, if they are not willing to renounce thought itself.

Born wrote to Wolfgang Pauli that Einstein had "got stuck in his metaphysics," and Pauli sadly agreed. Born said of Einstein's stance, "Many of us regard this as a tragedy—for him, as he gropes his way in loneliness, and for us who miss our leader and standard-bearer."

Nonphysicists sometimes joined in the critical chorus. During a bitter dispute with Einstein concerning the future of the Hebrew University in Jerusalem, Chaim Weizmann made the gratuitous and stinging remark, "Einstein acts like an aging prima donna who has lost her voice."

When he was seventy years old, Einstein wrote to his old friend from his Zurich days, Maurice Solovine, about his lifetime search for ultimate unity: "You imagine that I look back on my life's work with calm satisfaction. But from nearby it looks quite different. There is not a single concept of which I am convinced that it will stand firm, and I feel uncertain whether I am in general on the right track."

But when a person struggles to reach toward an ever higher goal, is not final failure inevitable? Said Eugene O'Neill, whose tragic drama won him the Nobel Prize for literature during Einstein's lifetime: "The people who succeed and do not push on to a greater failure are the spiritual middle-classers. Their stopping at success is the proof of their compromising insignificance. How petty their dreams must have been!"

SEVEN HUNDRED YEARS FROM NOW?

In the thirteenth century in Egypt lived Moses ben Maimon, better known as Maimonides, a great Jewish philosopher-theologian-physician-scientist who also sought to know the ways of the "Old One." Spinoza read Maimonides and was influenced by his views. In the winter evenings of 1951 Einstein read Maimonides and noted how his scientific ideas were limited by the knowledge of his time.

Seven hundred years from now, he mused, our present scientific thought would appear as strange to the people then living as the science of Maimonides' time appears to us now.

Today, in the digital computer age, we are beginning to explore the complexities that can arise from the digitized description of events. A new theory of fundamental reality underlying space and time may evolve that is based on digital—that is, discrete—elements.[2] Einstein was open to such possible development. He told Rudolf Carnap: "Physics must be one or the other, pure field physics with all magnitudes having continuous scales or else all magnitudes including space and time must be discrete."

But seven hundred years from now? In a letter to a friend Einstein wrote: "The mathematical God smiles . . . he knows how far we are from real depth of understanding."

---

2. Leading scientists in the new fields of cellular automata and complexity theory have conjectured that fundamental reality may be digital computer-like— that is, based on discrete elements.

Will Einstein Be Vindicated?

After Einstein died, theoretical physicists became more sympathetic with his view that quantum theory does not represent fundamental truth. Heralding this change in attitude in 1975, P. A. M. Dirac, also a Nobel prize–winner in physics and the last of the great pioneers of quantum theory, said of its status then:

> Our present quantum theory is very good provided we do not try to push it too far. . . . We do not try to apply it to particles with very high energies and we do not try to apply it to very small distances. When we do try to push it in these directions we get equations which do not have sensible solutions. We have our interaction always leading to infinities.
>
> . . . It is because of these difficulties that I feel the foundations of quantum mechanics have not yet been correctly established. Working with the present foundation people have done an awful lot of work in making application in which they can find rules for discarding the infinities but these rules, even though they may lead to results in agreement with observations, are artificial rules, and I just cannot accept that the present foundations are correct.

7

"I Could Not Feel So Close to Spinoza Were I Not Myself a Jew"

Spinozism is the religion of the future.
GEORG CHRISTOPH LICHTENBERG

"I could not feel so close to Spinoza were I not myself a Jew," Einstein wrote. His words are even more surprising than first appear. He knew that Spinoza, though born a Jew, took a scornful attitude toward the Jewish people and derided them in his writings. Einstein was usually quick to denounce those who indulged in such tactics. What does his remarkable statement tell us of his own beliefs?

He said it in replying to a letter commenting on the great difference between the detailed specificity of Jewish law and the cosmic nature of Spinoza's philosophy:

... it seems to me that you are right in saying that the gulf between Jewish theology and Spinoza cannot be bridged. However, it seems to me no less true to say that Spinoza's worldview is penetrated by the thought and way of feeling which is so characteristic of the living Jewish intelligence. I feel that I could not be so close to Spinoza were I not myself a Jew, and if I had not developed within a Jewish environment.

He and Spinoza were separated by three centuries of radical change. They lived in different cultures, at different stages of western man's intellectual evolution, under different political conditions. Nevertheless Einstein felt that they had similar ways of viewing the world, a similarity he credited to the "living Jewish intelligence" each possessed, a mental trait he seemed to believe existed independently of time and place.

To try to understand Einstein's comment, let us look first at Spinoza's underlying ideas concerning the nature of life and mortality, then at the worldview of the ancient Hebrews which we will see has assumptions similar to Spinoza's, and finally at Einstein's thought in which we find resemblances to both.

SPINOZA'S IDEAS

In 1656 Baruch Spinoza, the promising twenty-four-year-old son of a Dutch Jewish family active in the synagogue, was excommunicated by the Jews of Amsterdam for heresy. It was probably because he proclaimed that he did not believe in a personal God or in the literal truth of the Bible and refused to shut up about it. The Jewish community had been established only a generation before by Portuguese Jews fleeing the Catholic Inquisition. Not wanting to offend their Christian neighbors who had welcomed them to Holland, they were likely more disturbed by what Spinoza publicly said than by what he privately believed.

Embittered, Spinoza had little good to say about Jews and Judaism for the rest of his life. The excommunication still stands. In 1956, David Ben-Gurion, Israel's first premier—who, like many Jewish intellectuals of his day, admired Spinoza—tried to induce the Amsterdam Jewish community to lift it, but they refused.

Young Baruch Spinoza attended a Jewish school, where he learned Bible, Talmud, and the Hebrew language. He also studied the Jewish philosophers of the Middle Ages who tried to bring Jewish belief into accord with the secular knowledge of the time. A brilliant student, too brilliant for the comfort of his teachers, he was disturbed by the discrepancies he believed he found between what the Hebrew Bible actually said and what his teachers said it said.

Spinoza is commonly misunderstood for the same reason as is Einstein: His thought is based on a premise similar to one on which Einstein's ideas are based but is not implicitly accepted as a premise of thought by most people. Recall Einstein's words: "The four-dimensional space of the special theory of relativity is just as rigid and absolute as Newton's space." Einstein's four-dimensional space included time. Spinoza similarly asks us radically to transform our thought by looking upon the world from a view that transcends ourselves—to see the world not as events following each other in time, but as an encompassing spacetime unity. Once Einstein's concept of spacetime is internalized as an axiom of thought, Spinoza becomes easier to understand.

Spinoza called the all-encompassing reality of which spacetime is part, God. Spinoza's God has no anthropomorphic qualities but represents all being in its ultimate unity. Human beings, he says, are aware of only two attributes of God: thought and extension—in more familiar terms, mind and matter. For Spinoza's God, as for Einstein's theory of relativity, events don't happen; they exist.

All living matter, Spinoza says, strives to perpetuate itself. Among human beings this takes the form of an instinctive desire to preserve their own lives, to overcome death in some fashion. Since past, present, and future all coexist in the mind of God, death is overcome by seeking to share the mind of God, by seeking the eternal in life. He called this attitude "the intellectual love of God." It is accomplished by leading an ethical and moral life in awareness of that which is eternal—in awareness of the presence of God.

Eternity, then, is not achieved by living forever; it is independent of the flow of time. It is achieved by sharing the mind of God, and is reached to the extent that one's conscious thought is similar to

God's eternal thought. Spinoza writes: "He who has a body capable of a great many things, has a mind whose greatest part is eternal." "A great many things" refers to depth and extent of understanding. There is no reason, then, Spinoza says, for such a person to fear death. Eternity, he explains in a letter to his physician friend, Lodewijk Meyer, represents "the infinite enjoyment of existence."

Unfortunately, this is likely the most frequently misinterpreted aspect of Spinoza's thought—unfortunate because he considered it the culminating conclusion to his system. He discusses it in the last pages of his masterwork, the *Ethics*. To understand his words, consider the nature of human perception—that is, how a person becomes aware of the world outside his or her own body. Lichtenberg, with his knack for pungent phrasing that Einstein cherished, expressed it well:

> To say we perceive *external* objects is contradictory; it is impossible for a man to go outside himself. When we believe we are seeing objects we are seeing only ourselves. We can really perceive nothing in the world except ourselves and the changes that take place in us. It is likewise impossible for us to *feel* for others, as it is customary to say we do; we feel only for ourselves. The proposition sounds a harsh one, but it is not when it is correctly understood. We love neither father, nor mother, nor wife, nor child: what we love are the pleasant sensations they produce in us. . . . Nothing else is at all possible, and he who denies this proposition cannot have understood it.

Spinoza, then, is saying that though the human mind has its physical basis within the confines of the body, it ranges in thought outside the body; and it is in what occurs outside the body, though experienced from within it, that we can find consummation for our lives—we can experience the eternal if our experiences are compatible with God's thought. The truly wise man, Spinoza concludes, "never ceases to be and always enjoys satisfaction of mind."

Though Einstein used the word "God" often, Spinoza easily outdid him. God's self-awareness, says Spinoza, includes everyone's eternal being (and more). To extend one's awareness further outside one's body within the greater world of being—thus becoming more aware of the self-awareness of others so that one empathizes with them—is to come closer to the self-awareness of God, is to come closer to experiencing eternity. In Spinoza's curt language: "The mind is eternal in so far as it conceives things under the form of eternity."

Spinoza was well aware that these ideas of eternity are not easy ones to grasp: "If we attend to the common opinion of men we shall see that they are indeed conscious of the eternity of their mind, but they confuse eternity with duration, and attribute it to imagination or memory, which they believe remains after death."[1]

---

1. Spinoza's words in this section have been quoted from the *Ethics*, Part V, Propositions 31, 34, 39, and 42.

AN ANCIENT AWARENESS

Though Spinoza seemed to enjoy derogating the Jews—one doesn't easily forgive being publicly excoriated—he believed that his understanding of reality was like that of Jewish tradition. In writing to Henry Oldenburg, the secretary of England's newly formed Royal Society, he said grudgingly, "I would dare to say that I agree also with all the ancient Hebrews as far as it is possible to surmise from their traditions, even if these have become corrupt in many ways."

Spinoza surmised the thought of the ancient Hebrews from the Bible and the Hebrew language in which it was written. A language both is based on a view of reality and helps create it—and thus implicitly tells us of the worldview of those who speak it. Spinoza was aware of this but also aware that the Hebrew language has changed since biblical days—in fact, it was already different at the time of Jesus. He wrote that a history of the Hebrew language is indispensable in order to understand the Bible because all the authors of the Old and New Testaments were Hebrews. He started to write a Hebrew dictionary to facilitate this study, but never completed it.

The idea that reality is more than the present moment—that the past and the future are also real—is not new with either Einstein or Spinoza before him. It is the manner of thinking of biblical man and woman. The ancient Hebrew saw the past as laid out in front of him, the future as behind him—unseen but there.

Biblical Hebrew does not have tenses indicating past, present, and future as do modern languages. Instead of tenses it has two aspects, one indicating that action is incomplete, the other that action is completed. Biblical Hebrew blurs the distinction between past, present, and future. The prophets speak of future events as if they had already occurred. Jewish tradition tells us that all generations of Israel—past, present, future—were present at Sinai when Moses received the Ten Commandments.

It has been long noted that the Hebrew Bible does not talk about life after death. Three hundred years ago, Jacques Bossuet declared that it must be because God did not consider the ancient Hebrews intelligent enough to grasp the concept of immortality! But the reason is less gratuitously insulting to the people who had the genius to write the Psalms and the Book of Job.

It is the nature of life, as Spinoza points out, to seek to survive. The biblical man or woman, like today's man or woman, instinctively sought eternal life—survival in its highest form. But "eternal life" has a different meaning than limitless length of time to a people who do not consider the present moment to be ephemeral but part of a richer reality which includes past and future. Eternal life to the ancient Hebrew has nothing to do with life after death but is a function of the quality of life rather than its extent.

In one of the older parts of the Bible, the First Book of Samuel, this understanding is put primitively but dramatically (chapter 25, verse 29):

And should anyone rise up to pursue you and to seek your life, then your life shall be bound in the bundle of the living with YHVH your God, but the lives of your enemies he will sling out as from the hollow of a sling.

Hence, the biblical psalmist was less concerned with the temporal limit to life than he was with being separated from God during life. "You [God] cut off all who go whoring from you," says the Seventy-third Psalm, but then it sings: "As for me, nearness to God is my good."[2]

The eminent Danish biblical scholar, Johannes Pedersen, tells us in his magisterial study, *Israel, Its Life and Culture*, that the ancient Hebrews saw all generations—past, present, future—to be fused into a great whole, called *olam*, "eternity," from which human events spring. Biblical "eternity," he writes, "is not the sum of all the individual periods, nor even the sum with something added to it; it is 'time' without subdivision, that which lies behind it, and that which displays itself through all times."

English versions of the Bible translate *olam* as "eternity", "forever", "world", "lifetime", "remote

---

2. There is a line in the Seventy-third Psalm that reads in English translation: ". . . And afterward receive me to glory." This line is often cited as showing belief in life after death. But a leading Hebrew scholar who has studied the psalm in depth, Sheldon H. Blank, has pointed out that this is a mistranslation from an uncertain Hebrew text.

time", or other words, because no single English term corresponds to it. But such limited translations miss the essence of its meaning. To avoid being misleading, *olam* is best left untranslated.

The depth of participation in *olam* depends on closeness to God—closeness to the roots of reality. The extent of that closeness, says the Bible, especially in the Psalms when they speak of the presence of God in a person's life, is determined by the nature of the life he leads. Psalm 41, for example, in verse 12, reads:

> As for me, you uphold me because of my integrity,
> And you set me in your presence *olam*.

Psalm 37 is an acrostic, the first letter of each pair of verses spelling out the Hebrew alphabet. The arrangement served as an aid to memory, suggesting the importance of the psalm to biblical man and woman. Verses 18 and 27 say:

> The Lord knows the days of the blameless
> And their inheritance is *olam*.
>
> Depart from evil and do good,
> So that you dwell *olam*.

For the word's use in the Bible on a less elevated plane, turn to the First Book of Kings, first chapter. Bathsheba, knowing that her husband, King David, is close to death, entreats him to proclaim that their son, Solomon, will succeed him even though Solomon is not next in line of accession. David agrees, and Bathsheba exclaims, "May my Lord, King David, live *olam!*"

WHAT EINSTEIN BELIEVED

Einstein's own existential beliefs were close to Spinoza's. He considered them not to be unique, but to be the conclusions that would be arrived at by any probing, thinking person. "The more deeply philosophical doctrines as far as questions of existence are concerned are thought about," he wrote, "the less different they are from each other."

Though he praised Moses, the Buddha, and Jesus as mankind's greatest benefactors, with admirable impartiality he considered *all* of today's organized religions to be decadent—in fact, to have been in a state of decline since the days of the biblical prophets, of Jesus, and of Gautama. He believed that the religions they inspired have fallen into the hands of entrenched, self-serving hierarchies more concerned with maintaining power than in pursuing truth. "If one purges the Judaism of the prophets and Christianity as Jesus taught it of all subsequent additions, especially those of the priests," he wrote, "one is left with a teaching which is capable of curing all the ills of humanity."

"Christianity as Jesus taught it," he believed, was essentially "the Judaism of the prophets." A Catholic science student, concerned for Einstein's soul, wrote to him begging him to convert to Christianity: ". . . pray to God and his mother and see a Catholic priest as soon as possible." Einstein responded, "If I would follow your advice and Jesus could perceive it, he, as a Jewish teacher, surely would not approve of such behavior."

Einstein on Buddha, Moses, and Jesus

Our time is distinguished by wonderful achievements in the fields of scientific understanding and the technical application of those insights. Who would not be cheered by this? But let us not forget that knowledge and skills alone cannot lead humanity to a happy and dignified life. Humanity has every reason to place the proclaimers of high moral standards and values above the discoverers of objective truth. What humanity owes to personalities like Buddha, Moses and Jesus ranks for me higher than all the achievements of the enquiring and constructive mind.

What these blessed men have given us we must guard and try to keep alive with all our strength if humanity is not to lose its dignity, the security of its existence, and its joy in living.

But why did Einstein couple "Judaism of the prophets" and "Christianity as Jesus taught it" with the teachings of the Buddha?

What they do have in common is their concern with life: each sets forth a way of living that attempts to elevate the spirit of man. Ideas associated more with death than life—such as resurrection of the body, immortality of the soul, and transmigration of souls—entered Judaism, Christianity, and

Buddhism only with the blurring of the original visions of their founders. Neither Moses, Jesus, nor Gautama had anything to say about life after death. To each it was an alien concept with no place in his view of reality. Each would have agreed with Spinoza's words: "A free man thinks of nothing less than of death, and his wisdom is a meditation not of death but of life."

Einstein concurred. He believed that man's sense of self-ness, his "soul" so to speak, was an aspect of his physical being:

> The fact that man produces a concept "I" besides the totality of his mental and emotional experiences or perceptions does not prove that there must be any specific existence [reality] behind such a concept. We are succumbing to illusions produced by our self-created language, without reaching a better understanding of anything. Most of so-called philosophy is due to this kind of fallacy.

And when his brother-in-law, Paul Winteler, died—a year after Winteler's wife, Einstein's sister, Maja, had died after much suffering—Einstein wrote these words:

> It is a good thing that this individual life has an end with all its conflicts and problems. . . . Those who brought about the belief that the individual continues to live after death must have been very sorry people indeed.

He may have disdained those who refused to accept their mortality, but, despite his protests, like all men, he was concerned about his own. Worried about his health during his Berlin years, he selected as personal friends mostly Jewish physicians.

Einstein responded to the elemental survival instinct within him in his own way, akin to the biblical way. Consider his comment on a statement of Hugo Muensterberg's concerning immortality.

Muensterberg was a University of Berlin psychologist of Jewish birth who, later in his career, was invited by William James to join him at Harvard. Though Muensterberg, following a then well-trodden path for aspiring academics, had rejected his Jewish background, he formulated a concept of eternity as an aspect of living that appears to have developed from it.[3]

When he had been asked if he believed in the survival of personality after death, Muensterberg answered: "I cannot conceive of personality in terms of time." Was Muensterberg, an interviewer asked Einstein, evading the question with that response? "I do not think so," Einstein replied. "It was the only possible answer." But Einstein was not himself being evasive. He agreed with Muensterberg because his world was that of relativity theory's spacetime, a world frozen in four dimensions in which a person just is. Said Einstein, "A division of time into past, present, and future does not exist in physics."

---

3. He presented his ideas in a small book, *The Eternal Life* (Boston: Houghton, Mifflin, 1905).

"I am part of eternal being," Einstein affirmed. It was in this belief that Einstein's visceral desire to preserve his own consciousness was satisfied.

What is this "consciousness" about which all of us are so concerned and of which we speak so glibly? He reminds us that we, including himself, do not really know what we are talking about:

> We know consciousness as the essential part of our ego and by analogy as the essential part of other egos. The poverty of our expression does not show us more of it. We can only guess and even this guessing does not have a clear meaning to our thought. There seems to be no other attitude than humility and modesty. The only thing I am feeling strongly about is: It seems foolish to extend our personality beyond our life in both directions and we do not know what consciousness means outside the frame of personality.

Einstein's beliefs were tempered by the attitude of humility and modesty of which he speaks, by awareness of how little we know and how little we can know, by awe for the mystery of being and its unity—a mystery which he called God.

He more than once said that he believed morals and ethics to be humanity's most important problem, but that it was a problem for humanity and not for God. He said he believed in Spinoza's God, "who does not concern himself with the fate and actions of men."

But it is the nature of man, according to Spinoza, to concern himself with God. He would have added to Einstein's statement: Morality is not a problem for God but a problem for humanity in trying to achieve closeness to God.

WHEN THE BIBLE, SPINOZA, AND EINSTEIN ARE MISUNDERSTOOD

Einstein, Spinoza, and the Hebrew Bible have been commonly misunderstood when their view of reality—as ancient as the oration of the early prophets and as recent as the development of relativity theory—is not considered. Verse 26 in the nineteenth chapter of Job is often said to be one of the more inexplicable passages in the Hebrew Bible:

> Even after my skin is destroyed, yet from my flesh I shall see God.

Some scholarly critics say Spinoza must have abandoned his senses, contradicting his own thought, when he wrote in the *Ethics*:

> The human mind cannot be absolutely destroyed with the human body, but there is some part of it that remains eternal.

And Einstein's seemingly equivocal assertion on immortality, made when he was fifty-four years old, baffles those who come across it in his letters:

> I believe the mind is immortal in the same sense as the body for it is difficult to doubt that the capacity to build living bodies and consciousness is

connected with matter. But I see no justification to extend personality beyond the span of life of the individual.[4]

To better understand these statements—Job's, Spinoza's, Einstein's—consider Lichtenberg's comment on human perception of the world (quoted earlier in this chapter) in connection with your own outlook on reality. Though your personality is one with your body, you do not feel that your world of being is confined to your body. To do so would be to feel like an animal, and an animal of minimal consciousness at that. You get fulfillment as a human being only from your awareness of what occurs outside your body. A familiar example: a mother's contentment watching her child happily play across the room. The love you feel for another requires the existence of the other outside the spacetime bounds of your feeling body.

Job's, Spinoza's, and Einstein's declarations become clear when it is realized that for each the world is one of spacetime being and the physical body is the locus of an awareness which extends outside the body into spacetime—living past and living future included.

---

4. Notice the aging Einstein's change in tone from his words of a dozen years earlier: ". . . nor would I want to conceive of an individual that survives his physical death; let feeble souls, from fear or absurd egoism, cherish such thoughts. I am satisfied with the mystery of the eternity of life. . . ."

THE EINSTEIN–SPINOZA–JUDAISM
CONNECTION

The biblical prophet comforted his people with the promise of a golden age on earth occurring in the future. The people found consolation by making this coming age part of themselves. Since past, present, and future events were all part of their being, the Hebrew prophets emphasized the need to sanctify living itself, which created that past and will create that future. Einstein agreed:

> Judaism seems to me to be concerned almost exclusively with the moral attitude to life and in life. . . . The hallowing of the supra-individual life brings in its train a reverence for everything spiritual—a particularly characteristic feature of the Jewish tradition. . . . The sanctification of life in a supra-personal sense is demanded of the Jew.

He believed the Hebrew Bible to be the result, not the cause, of Jewish mentality. He writes:

> To me, the Torah and the Talmud are merely the most important evidence of the manner in which the Jewish concept of life held sway in earlier times.

He then poses the question, can this Jewish understanding of life "be found anywhere else under another name?" He replies:

> In its pure form, it is nowhere to be found, not even in Judaism, where the pure doctrine is ob-

scured by much worship of the letter. Yet Judaism seems to me one of its purest and most vigorous manifestations.

Einstein's theory of relativity, in which events past and events future are as real as events present, puts the old Hebrew understanding of reality, from which the Jewish view of life springs, in comprehensible terms for today—moreover shows this ancient view to be both more sophisticated and more scientifically advanced then our own.

While rejecting Jewish ritual and any idea of a chosen people as "superstition," he considered biblical thinking and Spinoza's thinking to be representative of Jewish thinking, a way of thought he shared. "The intellectual love of God is eternal," said Spinoza. Striving for it, said Einstein, embodies the ideals of Judaism:

> For myself the unadulterated Jewish religion is like all other religion an incarnation of primitive superstition, and the Jewish people to whom I gladly belong, and whose mentality I deeply share, has for me no different dignity than all other peoples. . . .

> . . . the idea of Judaism as I perceive it in an articulate way: a freeing from fossilized ritual, a clinging to the social ideal which was so early formulated, a fight against evil in every form, a striving for Spinoza's intellectual love for God in which the special and philosophical ideals of Judaism are united.

8

A Question of Faith

God created man in his own image, says the Bible, and the philosophers do just the opposite; they create God in theirs.

GEORG CHRISTOPH LICHTENBERG

When his lifelong friend Michele Besso died in Europe, Einstein sent his family a note of comfort that included these words:

> . . . And now he has preceded me briefly in bidding farewell to this strange world. This signifies nothing. For us believing physicists the distinction between past, present, and future is only an illusion, even if a stubborn one.

This message is among Einstein's last. He died a few weeks after penning it. The times he had spent with Michele Besso, he is saying, remain real in flowless spacetime. He had talked about human relationship existing in Spinoza-like eternity on other occasions:

> One has a feeling that one has a kind of home in this timeless community of human beings that strive for truth.

And he believed that Jesus thought in this manner:

> I have always believed that Jesus meant by the Kingdom of God the small group scattered all through time of intellectually and ethically valuable people.

If Einstein is right in this conjecture, then Jesus' understanding of space and time was the ancient Hebrew one.

You may remember that Einstein castigated Besso for believing that only the present moment is real, for not accepting the reality of events past and events future. But in writing to Besso's family is he now implying something more about the relation between individual awareness and spacetime? In earlier years, writing as a younger man, he did not give that relationship the very personal nature he now seemed to be doing. It is a relation that touches upon the mystery of consciousness and the "now," an enigma that he had told Rudolf Carnap he believed would always be beyond human understanding.

Kurt Gödel, who is to the field of mathematical logic what Einstein is to physics, was still living when I came across the letter to the Besso family in the Einstein archives in Princeton. Like Einstein, Gödel was a permanent member of the Institute for Advanced Study and the two often walked home together, absorbed in conversation, during Einstein's last years. Einstein once told Oskar Morgenstern that his own work no longer meant much, that he continued to come to the Institute mainly "to have the privilege to walk home with Gödel."

I knew that Gödel shared Einstein's view of time since he, in a paper favorably reviewed by Einstein, had accepted the reality of past and future. He does so by describing a hypothetical world which satisfies Einstein's equations for relativity in which, Gödel writes, "by making a round trip on a rocket ship in a sufficiently wide curve, it is possible . . . to travel into any region of the past, present, and future, and back again, exactly as it is possible in other worlds to travel to distant parts of space." It was undoubtedly the first time the possibility of time travel was put on firm mathematical footing. I later was told that Gödel was very concerned about his own mortality—to the extent of extreme reclusiveness coupled with an abnormal fear of life's risks.

I called Gödel, who was kind enough to speak to me,[1] and read to him Einstein's words to Michele Besso's family. What, I asked him, did Einstein mean? I was not at all prepared for the answer he gave me: After hesitating a moment, he said, "I

_____

1. I talked to Gödel on August 14, 1977—just five months before his death from self-starvation induced by a paranoid fear of food poisoning.

think he meant it as some sort of joke." Einstein, he continued, believed that science could never explain consciousness: "the methods of science lead away from the life-world." He then added that Einstein believed the concepts of space and time to be human construction. Reality was something deeper.

Einstein joked often. But in his archives I had read many letters of condolence that he had written, and he did not joke in them. I cannot believe that Einstein, knowing that he himself had not much longer to live, would write these words to the grieving family of his oldest friend in jest.

Gödel has reported that he and Einstein talked about philosophy, physics, and politics as they walked home together from the Institute for Advanced Study and that they often disagreed. The mathematician Hao Wang, who had many conversations with Gödel, says Gödel believed that mind and matter are separate, everything that happens has an ultimate purpose, and that God is a person. Clearly he and Einstein had much on which to argue.

In a letter to his mother, Gödel wrote that in their walks Einstein "kept pretty much to himself with respect to personal questions." We cannot know the specifics of their philosophical conversation, but I can guess the turmoil in Einstein's mind, be it at or below the level of conscious awareness, that may have led to Gödel's response to me. For there are terrible connotations to this belief in a living reality that spans spacetime.

If the good times in the past—times of friendship, of joy, of love—are real, are not the times of suffering in the past—times of disease, of tragedy, of agony in Nazi extermination camps—equally real? Wrote Einstein to Michele Besso, in the letter admonishing him to take seriously the four-dimensionality of spacetime reality: "You say that suffering is essential which physically interpreted is linked to events which cannot be changed, as, of course, all remembering. And in this I agree with you completely."

Relatives, among them the wife and young daughters of his cousin Roberto Einstein with whom he had grown up; colleagues, as early ones as Georg Pick and Emil Nohel with whom he had worked at the German University of Prague; and countless others who had shared his life suffered and perished as victims of the Nazis.

During those years of Hitler's terror Jewish refugees from Germany found the doors of most nations closed to them. The United States accepted very few and the requirements for admission were difficult to satisfy. One was that a potential immigrant needed an affidavit from a sponsor who would agree to provide financial support if needed. Nohel wrote to Einstein asking him to sponsor himself and his family for admission to America. But Einstein had provided so many affidavits he felt he could not do so and maintain his credibility. He turned the letter over to another for action, saying that he would help Nohel financially after he arrived in America. But, no action was taken. Emil Nohel and his family died in German concentration camps.

Einstein did come to offer to sponsor so many refugees that the Department of Immigration stopped paying attention to him. They knew that his income was inadequate to guarantee support for so many.

Roberto Einstein, the son of his father's brother and business partner, Jakob, was an engineer living in Italy, near Florence. On August 3, 1944, when Roberto was away, the Germans murdered his wife and their two young daughters in their home. Then they burned down the house. A year later, Roberto killed himself.

"I am part of eternal being," declared Einstein. Did he then believe that evil incarnate was also part of eternal being? If so, how could he do other than find unadulterated being-in-the-world starkly, intolerably, bleak?

They are painful questions, perhaps too painful for Einstein to have directly addressed, but we can infer his internal response from his correspondence in which he expressed himself more freely than he did in material meant for publication.

He often talked of Spinoza in his letters. He urged people to read Spinoza's writings—to seek out what he called "nuggets" within them. He asserted time and again that he did not believe in a personal God but he did believe in the God of Spinoza. In Spinoza's conception of God only the aspect of man's consciousness that is in accord with God's consciousness—in accord with ultimate unity —is eternal. It would include that which strengthens true human relationship: creative accomplishment which others can make part of themselves, love for others. But it would not include that which strives to destroy such relationship.

The Bible, on which Spinoza's thought is based, concurs that evil is not retained in the book of life. Says Exodus, chapter 32, verse 33: "And the Lord said to Moses, whoever has sinned against me, I will wipe him out of my book."

Therein we see the form an answer to these questions can take: the horrors of the Holocaust do not exist in the eternal, but the acts of the persecuted in mutual support, and their fierce desire to provide witness of their plight as a preventive for future generations, do. The thought is comforting—but Einstein wondered if Spinoza would have been able to maintain his beliefs in the face of Nazi evil. He wrote to a Jewish survivor:

> I was very sorry to hear in your letter that you have suffered such sad losses and that you are almost the only one in your family who has not become a victim of these murderers. When one sees how the rest of mankind treats us in view of this, one is filled with pervading disgust. I wonder if Spinoza would have found strength to rise above it without inner damage. It is good that he was spared this brutal test.

Einstein often wrote of people whose thought he respected, such as Spinoza, in words that applied to himself. Was he himself able "to rise above it without inner damage"? Was he still able to say with sincerity, "I believe in Spinoza's God"?

Time's Eye

The poet Paul Celan was victim, survivor, and victim again of the Holocaust. His parents perished in a German concentration camp. In 1970, in midcareer at the age of forty-nine, he drowned himself in the Seine. The translation from German is by Michael Hamburger.

This is time's eye:
It squints out
from under a seven-hued eyebrow.
Its lid is washed clean by fires,
its tear is hot steam.

Towards it the blind star flies
and melts at the eyelash that's hotter:
it's growing warm in the world
and the dead
burgeon and flower.

PAUL CELAN

9

Striving with a Stubborn God

The only way of venerating God is to do our duty and to act in accordance with the laws reason has given us. 'There is a God' can, in my view, mean nothing other than: with all my freedom of will, I feel myself compelled to do right.

GEORG CHRISTOPH LICHTENBERG

It was once pointed out to Einstein that he had just stated a view contradicting his words of a few weeks earlier. He retorted that the Good God did not care whether he was always consistent.

His response—like many of his remarks in which he manages to interject the word "God"—seems incongruous. But is it? Underlying his view of the world was the theory of relativity supported by his reading of Spinoza: past and future were as real as the present, all three existing together in seamless unity. This unity, unfathomable in its depth, was his "Good God," who—or which—was not anthropomorphic and so not "concerned" with his actions.

But Einstein's "Good God" is not passively in-effectual. To try to understand his thinking, we must go back to his childhood. He was a religious child who composed little poems in praise of God, sang songs to God. But at the age of twelve, he tells us, "through the reading of scientific books I soon reached the conviction that much in the stories of the Bible could not be true."

Disillusioned, he turned away from religion to seek salvation in science. When his theories of space, time, and spacetime triumphed over earlier understandings of the universe, he was sure he had taken the right path. In that pursuit of scientific knowledge, he wrote, "the introduction of the concept of a God seems to be without value."

But the God of the Hebrew Bible was a stubborn God. He had refused to be summarily dismissed from Spinoza's thought processes in an earlier century and, refusing to be abandoned now, remained in the recesses of Einstein's mind. There the biblical God waited to be rediscovered in the maturer perspective that added years would bring. This God, the Hebrew prophets proclaimed, wanted to be reached not by passive contemplation but by human beings' acts of concern for each other which reflected his will.

Einstein saw the terrible tribulations of the world about him and longed to escape them in science and philosophical reflection. But when humanity's plight in Europe worsened with the rise of Hitler, the silent, persistent urging of God within him would not let him. In a letter to an old friend, Queen Elizabeth of Belgium, whom he had met before leaving Europe for America, he confided his split feelings:

> The moral decline we are compelled to witness and the suffering it engenders are so oppressive that one cannot ignore them even for a moment. No matter how deeply one immerses oneself in work, a haunting feeling of inescapable tragedy persists.
>
> Still there are moments when one feels free from one's own identification with human limitations and inadequacies. At such moments, one imagines that one stands on some spot of a small planet, gazing in amazement at the cold yet profoundly moving beauty of the eternal, the unfathomable: life and death flow into one and there is neither evolution nor destiny; only being.

Striving with the biblical God within him, he at one moment would condemn the Bible as a book of primitive superstition, and at another praise it as the unquenchable source of the Jewish people's exalted ethical values. He would one moment declare that the word "God" should be abolished from use because of its personlike implications, and at the next use it in the very manner he opposed.

METAPHOR AND BEYOND

We have been speaking of God in metaphor. But Einstein has reminded us (see chapter 6) that our language concerning reality outside our own bodies is necessarily metaphor. We know the world only by the image of it formed in our minds. "Most mistakes in science and philosophy," he cautioned, "are made because the image is taken for the reality." He likened human understanding to a floating island adrift on the sea of truth, carrying its own premises on which its self-understanding is based. "A striving for a firmer basis," he said, "is a kind of hunt after ghosts." He of course was speaking in metaphor.

"Given to us," Einstein wrote, "are merely the data of our consciousness, and among these data only those form the material of science which allow univocal linguistic expression."

Reluctantly, he came to the conclusion that scientific knowledge was not sufficient to guide or explain human life. Science said nothing about the sanctity of life. It could explain neither humanity's need for transcendence nor his own restless search for ultimate transcending unity. It could not alleviate the deep loneliness of existence. It could not truly replace the warm God of his childhood. Einstein writes:

Science can satisfy only one aspect of our soul. . . . It cannot replace art in its striving for beauty, nor can it give to man that type of consolation which is given by religion to many.

Science can never bring about the benevolent and selflessly serving attitude which is so essential for the social values of the individual.

He now believed that human life represents mystery beyond science's ability ever to understand—beyond that which can be expressed as an object of human thought.

> In my opinion, one cannot speak of human life in an objective sense as one speaks for instance of the mass of the sun.

If human life cannot be spoken of objectively, certainly the concept "God," cannot. His way eased by childhood memories, he returned to a new understanding of the religion of his fathers.

> Judaism is not a creed: the Jewish God is simply a negation of superstition, an imaginary result of its elimination. . . . It is clear also that 'serving God' is equated with 'serving the living.' The best of the Jewish people, especially the prophets and Jesus, contended tirelessly for this.

He viewed the world as did the biblical prophets twenty-five centuries earlier. They too assumed the living world, including both past and future, to be a seamless whole—one consisting of meaningful interrelation of human events. Our life has meaning, Einstein said, even though we cannot comprehend what that meaning might be other than that it involves serving others.

> The man who regards his own life and that of his fellow creatures as meaningless is not only unhappy but hardly fit for life.

> One can feel the meaning of life but one cannot understand it through reasoning.

Wittgenstein writes that to believe life has meaning is to believe in God. Einstein advised an

aged cousin who had suffered through the Nazi
era:

> Enjoyment of the small things, today crying, to-
> morrow laughing. However it happens, accept it.
> The deepest meaning remains in the dark.

In the dark, but there.

Einstein Sounding Like a Biblical Prophet

Einstein spoke the way Jeremiah and Isaiah—who also lived in times of spiritual and political crisis—did two-and-a-half millennia earlier.

About those American Jews who put material values ahead of spiritual ones:

> The dance about the golden calf was not merely a legendary episode in the history of our fore-fathers—an episode that seems to me in its simplicity more innocent than that total adherence to material and selfish values threatening Judaism in our own day.

About the threatening cataclysm if people do not change the way they think:

> The unleashed power of the atom has changed everything save our modes of thinking and we thus drift toward unparalleled catastrophe.

About his belief—though that belief was sorely tried, sometimes to despair—that a better age is coming:

> I hold that mankind is approaching an era in which peace treaties will not only be recorded on paper but will also become inscribed in the hearts of men.

10

Redemption

I believe that man is in the last resort so free a being that his right to be *what he believes himself to be cannot be contested.*

GEORG CHRISTOPH LICHTENBERG

Albert Einstein was no Baruch Spinoza. His pleasures were more earthy, his life more worldly. Spinoza led an abstemious, almost monklike, existence. Einstein enjoyed the company of attractive women even into later years and had affairs with some. A sample result: Clifford Odets, the playwright husband of lovely young Luise Rainer, who attained fame for her role in the motion picture *The Good Earth*, was jealous of Einstein's relationship with her. Einstein's secretary, Helen Dukas, possessed a photograph of Luise Rainer looking up

admiringly at Einstein as they walk on a Long
Island beach in the summer of 1937.

As much as he revered Spinoza and treasured
his insight, Einstein believed that Spinoza's response to living in a world of suffering was too
elevated for an ordinary mortal. Spinoza's answer
was to change one's view of life (to consider it from
the standpoint of eternity) rather than to change
life itself. Einstein said of it:

> It may have been more or less a solution for him-
> self, but surely not for ordinary people. And his
> solution is ultimately very similar to renunciation
> of life, a solution similar to that of Buddhism, a
> solution of pessimism. There is also an optimis-
> tic attempt at a solution: transformation of the
> wilderness into a garden, but the chance of suc-
> cess does not seem encouraging if you take into
> account the long struggle for the goal with dubi-
> ous effect. Herbert Samuel, in an interesting book,
> *Belief and Action*, tries to defend that standpoint.

Herbert Samuel believed that though the presence of God (he uses the less emotionally loaded
term "Deity") is evident in the laws that govern the
universe, God does not intervene in its operation.
He writes that our own human effort can solve the
problems of humanity, can create a better world—
and when we strive to do so we feel that we are
doing the will of God.

Einstein knew Samuel well. He had been the
British-appointed High Commissioner for Palestine during Einstein's visit there. An English Jew
from a wealthy banking family, Sir Herbert Samuel

had distinguished himself as both politician and philosopher. After returning to England, he became leader of the Liberal party in parliament and president of the Royal Institute of Philosophy. Einstein had corresponded with him over the years, had stayed at his home in London, and had invited him in return to Princeton.

Sir Herbert Lewis First Viscount Samuel was a man of intellectual bent, deeply influenced by his orthodox Jewish background. He had accepted the Palestine appointment with high hopes for the Jewish pioneers, stating afterward: "Twice in the history of mankind, on that soil had been engendered spiritual and cultural movements of supreme value to humanity. I held the belief—which I still hold—that Palestine in these latter days, might possibly add a third—on however modest a scale—to the religious revolutions of four thousand years and two thousand years ago."

Einstein read Samuel's *Belief and Action* with great empathy and even provided an effusive blurb to help promote its sales.[1] Two years later he was still sufficiently stirred to recommend the book to a symposium on religion at the Jewish Theological Seminary in New York. (See the first chapter for Einstein's words to the symposium.)

---

1. When I asked Helen Dukas if Einstein sometimes exaggerated to help books or authors that he liked, she rebuffed me primly but firmly, saying that Professor Einstein never wrote anything he did not fully believe.

Why did Einstein, instinctively critical, respond with such unbridled enthusiasm? After all, Samuel's thesis is an integral tenet of normative Judaism: Man and woman are partners of God in creating a better world. But Samuel clarified social and economic considerations and laid out what he believed to be practical steps for its accomplishment. Einstein called him an optimist "whose optimism is not based on shallow faith but on rich experience in living and clear, accurate thinking." Samuel, Einstein said, "speaks as a doctor of the soul."

Samuel saw in working to create a better world the way to spiritual fulfillment, to finding meaning in life. He concludes the original edition of his book, which is the one Einstein read, with these words, in which he uses the word "workman" in referring to God as creator: "We feel that the soul of the workman streams through us, and our lives take on an added greatness."

"I read Herbert Samuel's book with rare fascination and agreement," Einstein wrote.

Samuel's God, who is the source of the good but who makes the world better only by inspiring humanity to do it, has antecedent in Jewish tradition. The Talmud (*Yalkut Shimoni*, section 455) says: "'You are my witnesses,' says the Lord. 'When you are my witnesses I am God; and when you are not my witnesses, I am not, as it were, God.'" An old Hasidic riddle asks: "Where does God dwell in the world?" The answer: "God dwells in the world where man lets him in."

Einstein concurred, emphasizing, during the struggle against the evil of Nazi Germany and its allies, the need for immediate action:

We cannot stand aside and let God do it. Whatever there is of God or goodness in the universe, it must work itself out and express itself through us. We have to act in our time and in the nation whose way of life we carry. What we do is of supreme importance to all humanity, to history, to human destiny.

Samuel on the Mind of God in Belief and Action

Science . . . now recognizes the incompleteness of its own presentation. Since it recognizes that there must be 'something else' it gives room for Deity. . . . It is only when we discern in nature itself the reign of law, and in the law the hand of God, that we may see a divine splendor in the natural that is about us, and may open an access to what lies beyond.

If the cosmos is the effect and God the cause, the nature of the cause must be seen, if only in part, in the nature of the effect. From the music we infer the musician, from the picture the painter, from the thought the thinker, and from the universe the Deity. "The more we understand individual objects," says Spinoza, "the more we understand God." And since there is mind in the cosmos there must be mind in the Deity.

Minds of the human order have will and purpose. It cannot be supposed that a mind of the cosmic, creative order, fundamentally different as it must be, would be without those qualities.

If we see Deity as origin, the sequence of events as its handiwork, the human personality as an integral part of the sequence, then we see ourselves both as parts of the work and participators in the working. We feel that the soul of the workman streams through us, and our lives take on an added greatness.

A GOD LESS REMOTE

As he aged, Einstein's concept of God seems to have become less remote from humanity. Peter Bucky, the son of one of his few intimate friends, grew up knowing Einstein and often chauffeured him—Einstein never owned a car or learned to drive. He writes that Einstein "composed a number of songs to honor God, which I heard him sing to himself many times. I also heard him say that anybody who loves nature must love God. He also told me once that ideas, as such, stemmed from God."

Einstein called his attitude toward the word "God" one of "humility corresponding to the weakness of our intellectual understanding of nature and our own being." After his disillusion with science, he associated goodness with that mystery: "the passionate will to contribute to the improvement of human conditions must come from an independent source."

LIVING IN A WORLD OF SUFFERING

With his reputation for being both kindly and wise, Einstein received a continual stream of letters in Princeton asking for his views or advice. He tried to answer the most sensible and most needful ones. To simplify responding, he would use the same carefully worded reply in answering the same questions.

Sometimes these letters were from families in pain who had suffered tragic personal losses and were asking for words of consolation. He would reply with this letter, which touches on the essence of his religious attitude, on what he called the "transcendental interpretation of life":

> A human being is a part of the whole, called by us "Universe," a part limited in time and space. He experiences himself, his thoughts and feelings as something separated from the rest—a kind of optical illusion of his consciousness. This delusion is a kind of prison for us, restricting us to our personal desires and to affection for a few persons nearest to us. Our task must be to free ourselves from this prison by widening our circle of compassion to embrace all living creatures and the whole nature in its beauty. Nobody is able to achieve this completely but the striving for such achievement is in itself a part of the liberation and a foundation for inner security.

In talking to the journalist Raymond Gram Swing, Einstein put the thrust of that way of thinking into rawest terms: "I think the highest human achievement is for a man to put himself under the skin of another man so as to weep with him in his sorrow and rejoice with him in his joy."

Relief from the suffering of personal loss, Einstein is stating, comes not from the constriction of concern to exclude others, but its opposite: the expansion of concern, of selfhood itself, to include others. He then says that striving to do so provides a spiritual basis for living.

There is biblical precedent for Einstein's idea of the expansion of selfhood as a response to suf-

fering. The Hebrew word in the Bible that is translated as "salvation" connotes enlarging the space
of living and acting. The root of the word means
spaciousness. Psalm 118:5 reads:

> Out of my straits I called upon the Lord;
> He answered me with great enlargement.

And 1 Kings 4 tells us that God gave Solomon
"largeness of heart, like the sand that is on the
seashore."

William James in discussing the psychology of
religious experience says a person's consciousness
is "continuous with a wider self through which
saving experiences can come."

Einstein's own path to a wider self was through
his Jewishness. He expanded his span of passionate concern to include the Jewish people with all
humanity in the offing. He said to his fellow Jews,
"We strive to be one in suffering and in the effort
to achieve a better human society, that society
which our prophets have so clearly and forcibly set
before us as a goal."

The expanding circle of compassion of which
Einstein speaks then is not the passive compassion
found in eastern religion which accepts the state
of the world, but the active compassion of the biblical prophet who protests against the social evils
of his day. Einstein certainly did in his.

Locally, he complained that no blacks were
being admitted to the Institute for Advanced
Study—the reason, he said, being the undue influence of Princeton University, which would not ap-

prove. Nationally, associating himself with campaign after campaign to right social injustice, his words are often ringing. On the twentieth anniversary of the unjust execution of Sacco and Vanzetti in 1927, Einstein wrote:

> True justice cannot exist unless the people themselves are determined to be just; unless they practice brotherhood, respect the truth, and have the courage to resist blind prejudice and political passion.

"Whoever shuts his eyes to avoid seeing the bitter injustices of our times," Einstein said, "shares the guilt for their tragic continuation." His words echo those of the Talmud two thousand years earlier: "He who can protest and does not is an accomplice in the act."

"Man can find meaning in life, short and perilous as it is," Einstein summarized in his later years, "only through devoting himself to society."

FLIGHT FROM THE "I" AND THE "WE" TO THE "IT"

One of the many refugees from German terror that Einstein helped establish in America was the Austrian Jewish-born author, Hermann Broch. With no permanent place to live in the first years after his arrival, he stayed for a month—while Einstein was away—in Einstein's Princeton home.

Broch, too, believed that life's meaning lay in making a contribution to human society. Textile engineer by profession; inventor of a cotton mix machine, which he patented; and assistant director of his father's thriving textile business, he had achieved material success by midlife—but at the expense of his soul. Unhappy in work and marriage, he gave up both, divorcing his wife and then, when over forty years old, studying philosophy and mathematics at the University of Vienna. Becoming aware of his mortality, he wrote to his son, "All values we strive for are somehow connected with eternity, that is, with approaching death."

In the following years he wrote a series of acclaimed thought-filled novels that addressed moral and social questions. The critic George Steiner calls him "the greatest novelist European literature has produced since Joyce."

One of his themes was that science, in making our understanding of reality increasingly abstract, causes a breakdown of values and induces fear and anxiety. Influenced by Einstein's thought, he saw the theory of relativity as making possible an eventual reintegration of human values. In his later years he sought to develop a philosophical psychology that would provide humanity an "earthly absolute" as comfort and guide. His "earthly absolute" included overcoming the fear of death by denying the reality of the flow of time.

When he was twenty-three, Broch had converted to Catholicism—the majority religion in Austria. Now, nearing his life's end, he talked of converting back to Judaism. He spent much of his limited energy—he was not well in his last years—in helping newly arrived Jewish refugees from

Europe. But he was not what is considered saintly: Like Einstein, he enjoyed the company of attractive women; unlike Einstein, his liaisons were many and fairly public.

Though Einstein read widely in his search for understanding, in later life he said he had given up reading novels for lack of time. But he did read and was deeply moved by the novel Broch had begun in a German concentration camp and continued to work on when staying in Einstein's home. It was Broch's most ambitious novel and acknowledged to be his masterpiece: *The Death of Virgil.*

The book's story takes place in a time of intellectual crisis similar to Spinoza's and to our own: an age in which the old gods had failed. Virgil lived during the first century BCE, when Roman paganism was crumbling. Broch writes in great detail of a dying Virgil who wants to destroy the uncompleted manuscript of the *Aeneid*, on which he had been laboring for ten years.

With terrible clarity, Virgil sees in the last hours of his life that his work represented an effort to achieve only objective beauty and would be used to glorify the continuing oppressive political regime—it did not reach for the subjective truths that, by disclosing the divine in life, can alleviate human suffering. Broch wrote of the "supreme human compulsion" to "find a higher expression of earthly immediacy in the beyond," and to "lift the earthly happening over and beyond its this-sidedness to a still higher symbol." In destroying his own work of false art which did not point the way to that higher symbol, Virgil saw redemption for himself and removal of an obstacle toward redemption for humanity.

Einstein's science was his art. Beauty was one of its prime criteria. In his scientific work he saw expansion of his own being through creative accomplishment—accomplishment that was both part of himself and outside himself. But did it point any higher? Would it lead to the true "secret of the Old One"?

In reading Broch's book he was uncomfortably stirred as he looked back on his own life. He had escaped into the objective world of science, had not deeply committed himself to others, had not risked the possible pain of intimate personal relationship. Instead of persons, he had dealt with issues: social justice, peace between nations, the fate of the Jewish people. He maintained relationships with people at safe arms-length through correspondence.

Even his attitude toward most colleagues at the Institute for Advanced Study was detached. Herman Goldstine, who was engaged in designing and building a digital computer—in those days a difficult task that could result only in a massive vacuum tube machine that, though erratic and unreliable, was also exciting and promising—had an office next to Einstein's. He recalled, speaking intensely, that Einstein did not associate with him and showed no interest in his work. Nor did he see others go in and out of Einstein's office except his assistant of the time. He stated that the later careers of Einstein's assistants seriously suffered because in working for him they were removed from the mainstream of physics. Clearly he had felt hurt, offended even in distant memory, by Einstein's aloofness. Einstein, in his quest to know the ways of God, had no time for computers or their makers.[2]

His relationship with his sons was strained, a lifetime source of unhappiness for him. He blamed their mother, Mileva, who remained embittered. Eduard, mentally ill, was living in Zurich, mostly in an institution. Einstein, though concerned with his care, never crossed the Atlantic to visit him. Hans Albert taught civil engineering at Berkeley. Einstein became angered at him for leaving Judaism and embracing Christian Science at the behest of his wife—a step contributing to the tragic death of one of Hans Albert's sons, deprived of medical care.

In his mature writings we often find the words "compassion," "beauty," "truth." We rarely find

---

2. From today's vantage point he may have erred: complexity theory, which has evolved from computer theory, may provide a clearer avenue toward understanding the God of Spinoza than does unified field theory.

Complexity theory shows the mathematical consequences of relating materially distinct active components to each other in a complex network. But the living world is made up of just such dynamic components—namely, living selves. Spinoza's God represents the unity of all the disparate interrelated elements of the universe, including living selves.

the word "love." It had not always been that way. As a youth and as a young man Einstein had written ardent love letters. He wrote them to the pretty teenaged daughter of his teacher at the Swiss school he attended in preparation for college. He wrote them to his first wife before they were married. He wrote them to his second wife before they were married but while he was still married to his first. In each marriage, the loving words seem to have ended shortly after wedlock had begun. Both marriages, Einstein sadly admitted, were failures.

In a letter to young Peter Bucky, who was dating a girl of whom his mother disapproved, Einstein concluded with a sentence that suggests his own problem with marital fidelity. He wrote in German doggerel: "The upper half plans and thinks, while the lower half determines our fate."

Musing on his own life and on the import of *The Death of Virgil*, Einstein wrote to Broch:

> I am fascinated by your Virgil—and am steadfastly resisting him. The book shows me clearly what I fled from when I sold myself body and soul to science—the flight from the "I" and the "We" to the "It."

THE STREAM OF CREATION IN WHICH THE ETERNAL RESTS

Samuel's *Belief and Action* and Broch's *Death of Virgil* are very different books—one down-to-earth and simply written, the other lyrical, soaring to heights of poetic expression—yet the responses they invite to the question of how to live in a grievously imperfect world are not dissimilar. And they point to the answer Einstein found for a flawed man in a world of spacetime being: Transcend the often harsh and distressing facts of our own tiny share in the being of the universe by breaking through the wall of the self and growing beyond it. To bring to bear the sufferings, the hopes, the feelings of others, to strive to build a better world for and of humanity, is to experience life's meaning.

Since the world's yesterdays and tomorrows are real, expanding the boundaries of selfhood can extend it into past and future. The ritual of the Jewish Passover seder, in which participants declare "we were all slaves in Egypt," is a living remnant of the biblical belief in reality of the past and extension of selfhood over time. Albert Einstein restored that ancient understanding of time to us, made it acceptable for modern consciousness.

In binding ourselves to others in the past, in concern for others in the present, in building for others in the future, lie redemption. For in a universe-that-is, a universe of past-present-future isness, to do these things is to be a participant in the creation of a redeemed world and, in so doing, to become part of it.

"It is not up to you to complete the work," said the Talmud two thousand years ago, "but neither are you free to abstain from it." "Nobody can do much," said Einstein in our own time, "but the little one can do, one must do under all circumstances and at each moment."

Less than a month before he died, he sent Kurt Blumenthal, the man who brought him into the Zionist fold in Berlin thirty-five years earlier, this note:

> I thank you, even at this late hour, for having helped me become aware of my Jewish soul.

Hermann Broch writes of Virgil's final vision in the concluding paragraphs of his novel:

> . . . the immensity of the here and now, looking backward and forward at once, listening simultaneously to what was behind and what was ahead, and the rustling of the past, sunken into the forgotten invisibility, was rising up again to the present moment and became the simultaneous stream of creation in which the eternal rests. . . .

A Rumbling: truth
itself has appeared
among humankind
in the very thick of their
flurrying metaphors.

Paul Celan

(*translated by Michael Hamburger*)

Afterword
"Don't, Please, Compare Me to Gandhi"

A man has virtues enough if, on account of them, he deserves forgiveness for his faults.
GEORG CHRISTOPH LICHTENBERG

In 1983, when Jamie Sayen asked me if I would like to meet Albert Einstein's stepdaughter, Margot, and see the inside of his home I replied with an immediate yes. Jamie, the author of *Einstein in America*, had grown up in the house next to Einstein's on Mercer Street in Princeton. He often called on Margot Einstein, then in her eighties, to see how she was faring and to offer his assistance. She had been living alone in the house since the death of Helen Dukas a few years earlier.

Margot, a soft-spoken, gentle, white-haired woman, welcomed us at the door and gave Jamie permission to show me the house. The home looks much as it must have looked during Einstein's lifetime. The furniture he and his wife brought from their Berlin apartment, books, decorations, pictures, remain as they were. The bookcases include many volumes on religious topics, especially Jewish ones.

Upstairs in his study, bookcases line the walls, except for a large picture window overlooking a long garden behind the house. Among crowded books and papers, four pictures hang—pictures that Einstein must have greatly valued, for they occupy space that could have been usefully occupied by more books and papers. On one wall are portraits of the two great physicists who prepared the way for Einstein's work: James Clerk Maxwell and Michael Faraday. Faraday developed the fundamental ideas of field theory and Clerk Maxwell put those ideas into mathematical form. "Without their work," said Einstein, "there would probably have been no special theory of relativity and certainly no general theory of relativity." It is not surprising to see the two pictures there.

But it is surprising to see the only other pictures in that book-packed study, both hanging on the opposite wall. One is a portrait of a bearded rabbi, looking wise and compassionate. It is dated March 14, 1919—a gift from the artist, Margot Einstein tells me. That date was Einstein's fortieth birthday. Directly below the rabbi is another portrait, a highly stylized ink drawing of Mahatma Gandhi. His head is black, and white rays appear to be emanating from it, giving him the look of a saint. It is dated exactly thirty years later.

Only the rabbinic portrait is not of a specific person. Remembering Einstein's comment that if he were born in eastern Europe he probably would have become a rabbi, and his growing attachment to Jewish ways as he grew older, one wonders if he saw himself in it. But why the striking picture of Gandhi?

WHY GANDHI?

When Gandhi's autobiography was published, Einstein not only read it, he read it aloud to his family. He called the book "one of the greatest testimonies of true human greatness." It was entitled *The Story of My Experiments with Truth*. Gandhi, the man who sought God, used the word "Truth" often. Einstein, the man who sought truth, used the word "God" often.

As Gandhi tells in his autobiography, he discovered his life's vocation while practicing law in South Africa at the turn of the century. His first goal was to have the South African Indians treated more like whites and less like the badly oppressed blacks; his technique was moral appeal. The moral ambiguity of his lack of concern for the blacks—an ambiguity which made moral appeal incongruous—does not seem to have occurred to him or others at the time.

"In South Africa I was surrounded by Jews," remarked Gandhi about these years. Supporting him were three young non-Indians—all Jewish—

who would become his closest associates in the struggle: Henry Polak, Hermann Kallenbach, and his secretary, Sonia Schlesin.

Polak had joined Gandhi when he was twenty-two years old, shortly after emigrating from England. From a rabbinic family, he saw how similar discrimination against Indians was to discrimination against Jews, exclaiming ". . . this was the Jewish problem all over again!" He said he was acting "as a Jew who tries to remember that Judaism is a matter not only of belief but also of action."

At the time he met Gandhi, Hermann Kallenbach, a wealthy architect, loved luxury and spent extravagantly. Successful but not content, he was spiritually striving beneath his flamboyant exterior. At their first meeting Kallenbach wanted to talk about religion, and they discussed the Buddha's renunciation of life. Life-loving Kallenbach found Gandhi's ideas more palatable than those of the life-denying Buddha, and they soon became close friends.

When Gandhi's office needed a secretary, Kallenbach recommended Sonia Schlesin, an eighteen-year-old Jewish girl from Scotland. Enthusiastic, bright, highly idealistic, she took over the daily running of the office, refusing to accept anything more than a minimum wage. Her responsibilities grew through the years until she was acting as treasurer for the Indian rights movement, approving editorials for their newspaper, and advising Indian leaders on policy.

"My Dear Henry . . ."

Gandhi appreciated the incongruity of his closest collaborators being not Indians but Jews. While in England in 1909, seeking support from the British government, he wrote many letters to Polak in South Africa. On August sixth he wrote this one:

> This is how Mr. Dallow refers to you in his letter to the "Yorkshire Daily Observer": "Finding that all attempts to move the Imperial Government on grounds of justice to redress their grievances have failed, the Indian leaders have despatched one of their white sympathizers to India in the hope thereby of awakening the attention of the Indian people to their sufferings. The gentleman is an English Jew; an attorney by profession; in thought and habit a Hindu . . ." From one point of view, what a libel that you should be considered in thought and habit a Hindu. What would Kallenbach say to this? And yet from another standpoint, it is undoubtedly a compliment. You may regard it as neither.

SATYAGRAHA

Gandhi went to jail many times—in fact, seemed to enjoy it. He discovered that civil disobedience, in which an oppressive law is purposely broken en masse and the resulting jail sentence then willingly accepted, to be a very effective tool in a campaign for that which is morally right. It was one

of the stratagems for bringing about change without violence that Gandhi painstakingly fashioned
and refined over the years in South Africa. He
proudly named his methods "Satyagraha," or "Soul-
force." He explained it this way:

> For instance the government of the day has passed
> a law which is applicable to me. I do not like it. If
> by using violence I force the government to repeal
> the law, I am employing what may be termed body-
> force. If I do not obey the law and accept the pen-
> alty for its breach, I use soul-force. It involves sac-
> rifice of self.

Other tactics included demonstrations and
marches both to solidify the determination of the
protesters and to arouse sympathy from the general
public. We are now familiar with such techniques,
but it was Mohandas Gandhi in South Africa who
pioneered them and showed how effective they
could be even before the invention of television.
After their success in South Africa, Gandhi took his
methods to India, where he was popularly honored
by being called "Mahatma."

If all other measures for swaying the government failed, Gandhi found that refusing to eat
often worked. He was supremely confident that au-
thorities would eventually give in rather than let
him make a martyr of himself by starving to death.
Fortunately for him, he was always right.

Satyagraha was based on replacing mutual fear
with first mutual respect and then mutual trust.
Gandhi made the potent and remarkable discov-
ery that saintliness could also be shrewdness. One

effective stratagem was to consider sincerely the welfare of his opponents: When he was leading one Indian protest, white railway workers called a strike of their own. Railways were the essential backbone of the South African transportation system. Gandhi then halted his campaign and told the Indian laborers to go back to work, saying he would not take advantage of the government's problems by making conditions worse.

GANDHI AND ZIONISM

Though many of Gandhi's supporters were Jewish, Polak was critical of his fellow Jews for not backing Gandhi officially: "If the Jew does not stand up eminently for ethical principles, what *raison-d'être* has he in the scheme of things . . . for what purpose is his the Chosen Race?" But Kallenbach defended the Jewish community, saying that Gandhi did not expect organized Jewish sponsorship.

Not only did he not expect it, but Gandhi marveled that Polak, Kallenbach, and his other Jewish followers were acting so selflessly. They had nothing to gain for themselves from success of the Indian rights movement.

Did their altruism affect Gandhi's views on their own people's struggle, the Zionist effort to establish a Jewish home in Palestine? It did not. By then Gandhi had moved from South Africa to India in order to work for India's independence from Britain. Shortly before leaving, he had written to Kallenbach, "you still remain the dearest and nearest to me." Kallenbach journeyed to India to ask for his backing, but Gandhi refused to support the Jewish homeland. He would not risk alienation of Indian Moslems whose support he needed.

There is evidence of this in the Einstein archives: a letter to Einstein in 1947 from Jawaharlal Nehru, India's prime minister. At the time Nehru wrote, he and Gandhi were succeeding in their struggle with the British but were now confronted with savage fighting between Hindu and Moslem in the wake of the British agreement to leave India.

Einstein had sent an impassioned letter to Nehru asking his support for the settling of Jews in Palestine and concluding with these words: "I trust that you, who so boldly have struggled for freedom and justice, will place your great influence on behalf of the claim for justice made by the people who so long and so dreadfully have suffered from its denial."

In his reply Nehru said he had shown Einstein's letter to Gandhi. He expressed sympathy for the Jews and horror of the Nazis, but then he refused to give his support, giving Einstein a little lesson in international relations:

> As you know, national policies are unfortunately selfish policies. Each country thinks of its own interest first and then of other interests. If it so happens that some international policy fits in with the national policy of the country, then that nation uses brave language about international betterment. But as soon as that international policy seems to run counter to national interests or selfishness, then a host of reasons are found not to follow that international policy.

"DON'T, PLEASE, COMPARE ME TO GANDHI"

Einstein's and Gandhi's deeds and goals paralleled each other in remarkable ways. I am going to tell of them, an endeavor of which Einstein would not have approved: He met Gaganvihari Mehta, the Indian ambassador to the United States, some years after Gandhi's assassination by a Hindu fanatic, and Einstein commented to Mehta, "Don't, please, compare me to Gandhi. Gandhi did so much for humanity. What have I done? It is true I have discovered some scientific formulas but so have many other scientists. There is nothing unusual in that!"

Each lived simply, dressed simply—though Einstein did wear more clothes than Gandhi. Each laughed easily, especially at himself. Gandhi once wrote, "If I had no sense of humor I should long ago have committed suicide." When he visited King George V at Buckingham Palace, he was wearing only loincloth and sandals. "It hardly mattered," he declared; "the king had enough on for both of us." In humor, Einstein said, he found "consolation in the face of life's hardships."

In marriage, neither Einstein nor Gandhi seems to have known mature love—neither its ecstasies nor its quiet joys, neither its vulnerabilities nor its pains.

Each reacted to virulent racial prejudice by identifying with his own people's struggle for dignified survival and self-respect.

Each extolled simple living for his people. Gandhi wanted to establish self-sufficient agricultural villages in India. Einstein favored the agricultural *kibbutzim* Jews were establishing in Palestine. He wrote: "In Palestine it is not our aim to create another people of city-dwellers leading the same life as in the European cities and possessing the European bourgeois standards and conception. We aim at creating a people of workers, at creating the Jewish village."

In India large numbers, especially Untouchables, lived in misery, degradation, and sometimes starvation. Gandhi blamed the British, the rulers of India, for not doing more to alleviate conditions and also for fomenting hatred between the Hindus and Moslems of India in order to prevent them from uniting against Britain. But Gandhi respected the British and sought to change their policy by peaceful means.

In Europe Jews were being confined in concentration camps, were being slaughtered by the Germans. Einstein blamed the British, the rulers of Palestine, for not admitting more Jewish refugees and also for fomenting hatred between the Jews and Moslems in Palestine in order to prevent them from uniting against Britain. But Einstein respected the British and tried to change their policy by peaceful means. He called the terrorist activities of Menachem Begin's Irgun Zvi Leumi "a disaster."

Both Gandhi and Einstein journeyed to London to plead the cause of his people. Each was unsuccessful.

Each strove for peaceful relations between his own people and the Moslems who shared the land. Each was denounced for doing so; each was destined to fail. Now we sadly know how farsighted both were.

Gandhi, like Einstein, was aware of his own people's shortcomings. Appalled by the filth in India, by the primitive sense of basic sanitation, Gandhi wrote, "If we approach any village, the very first thing we encounter is the dunghill. If a traveler who is unfamiliar with these parts comes across this state of affairs, he will not be able to differentiate between the dunghill and the residential part. As a matter of fact, there is not much of a difference between the two."

Each was distressed by the lack of pride he often observed in his people—by the urge to pass, to assimilate within the majority culture.

When an American Jewish organization was formed to oppose Zionism and challenge the idea of Jewish peoplehood, Einstein reacted angrily:

> This organization appears to me to be nothing more than a pitiable attempt to obtain favors and toleration from our enemies by betraying true Jewish ideals and by mimicking those who claim to stand for one hundred percent Americanism. Our opponents are bound to view it with disdain and even with contempt, and in my opinion, justly. He who is untrue to his own cause cannot command the respect of others.

Einstein and Gandhi each called for courage in facing life. Fear, both said, had no place even in religious belief: "Where there is fear, there is no religion," wrote Gandhi. "A creed based on desires or fears merits no confidence," wrote Einstein.

To his fellow Jews, he declared: "This life is a precarious and unstable situation for all the living;

let us not therefore fall prey to a paralyzing pessimism, but let us affirm our existence courageously as did our fathers even in the days of greatest peril."[1]

Gandhi called courage a necessary component of Satyagraha: "What do you think? Where is courage required—in blowing others to pieces from behind a cannon, or with a smiling face to approach a cannon and be blown to pieces? . . . Believe me that a man devoid of courage and manhood can never be a passive resister."

A DIFFERENT IDEA OF REDEMPTION

But there is one vast difference between Einstein's thought and Gandhi's. They disagreed on the nature of fundamental reality and of human redemption.

Einstein wanted to sanctify the world of living humanity; Gandhi wanted to escape from it. In the preface to his autobiography, he writes: "What I want to achieve—what I have been striving and pining to achieve these thirty years—is self-realization, to see God face-to-face, to attain 'Moksha.'" Since he saw redemption as lying outside the world in which we live, individual human life, with the

---

1. From an undelivered address for a Brandeis University dinner. It was not delivered because Einstein withdrew when he learned that Cardinal Spellman of New York, whom he considered anti-Semitic, was also invited to speak.

possible exception of his own, was not sacred to him. He did not expand the horizon of his own selfhood to include other living selves.

During one of his sessions in a South African jail—in Volksrust—his wife became severely ill. He wrote to her that if she should die it would be a worthwhile sacrifice to the cause of passive resistance.

This is his advice to the Jews trapped in Germany during the Nazi terror:

> If I were a Jew and were born in Germany and earned my livelihood there, I would claim Germany as my home even as the tallest Gentile may, and challenge him to shoot me or cast me in the dungeon. . . . The calculated violence of Hitler may even result in a general massacre of the Jews. . . . But if the Jewish mind could be prepared for voluntary suffering, even the massacre I have imagined could be turned into a day of thanksgiving and joy that Jehovah had wrought deliverance of the race even at the hands of the tyrant. For to the God-fearing, death has no terror. It is a joyful sleep to be followed by a waking that would be all the more refreshing for the long sleep.

Gandhi's advice to the beleaguered British was no more palatable. In 1940 he said:

> Invite Herr Hitler and Signor Mussolini to take what they want of the countries you call your possessions. Let them take possession of your beautiful island. If these gentlemen choose to occupy your homes, you will vacate them. If they do not

give you free passage out, you will allow yourself, man, woman and child, to be slaughtered. . . .

He wrote in his autobiography, "To my mind the life of a lamb is no less precious than that of a human being." No belief could be further from the Judaism of Einstein or of Gandhi's Jewish friends.

Polak and Kallenbach submerged their personalities under his until he left South Africa. Then each reasserted his own individuality and the values of his own background. Polak returned to England, where he practiced law and raised a family. Kallenbach, who continued to practice architecture in Johannesburg, expressed a desire when he was in his sixties to live on a Palestinian kibbutz. He willed the bulk of his money to the then-new state of Israel.

Why then did Einstein hang Mahatma Gandhi's portrait in his study?

Gandhi was assassinated by Hindu religious fanatics in 1948. In the following year, Nehru—still at India's helm—came to Mercer Street to visit Einstein. With him was his daughter, Indira, who would later become prime minister. (It was a visit that Margot Einstein still recalled with pleasure when I spoke to her over thirty years later; she remarked in looking at Gandhi's picture that Nehru and his companions were more cultured than visitors from Europe.) In talking to Nehru, Einstein made two corresponding lists on a sheet of paper: in one he showed the development of the nuclear bomb; in the other, the parallel development of Satyagraha.

Einstein had signed the letter to President Roosevelt that led to the formation of that bomb and to the climactic end to the Second World War. Since then, feeling what he called "an overwhelming responsibility," he had striven to eliminate war between nations, had dreamed of a world where all lived in harmony. But he now feared that he had dreamed an impossible dream for the most basic of reasons, man's innate nature: "It is the fate of men that they are forced by congenital overpowering drives to make life hell for each other."

He admired Gandhi not for his alarming thought but for his accomplishment: the creation and implementation of Satyagraha, the peaceful resolution of conflict between peoples. He called it "by far the greatest achievement in the political field in the last centuries—not only for India but for the whole of humanity."

To Einstein, yesterday, today, and tomorrow were one. In Gandhi's methods he saw hope for tomorrow.

Appendix 1

CONCERNING THE EPIGRAPHS

Georg Christoph Lichtenberg, who is quoted at the head of each chapter, deserves to be better known. He lived in eighteenth-century Germany, but his writings were extolled by Tolstoy and Kierkegaard in the nineteenth century, by Einstein and Wittgenstein in the twentieth.

Lichtenberg was professor of physics, astronomy, and mathematics at the University of Goettingen. A statue of him, sitting on a bench with legs crossed and his hand on a book, stands—or rather sits—in a quadrangle there. Those who feel that today's standards for college admission have fallen too low will be heartened to learn that the situation was no better two centuries ago in Germany. Lichtenberg remarked, "It is incredible how ignorant students are when they come to the University."

He was a polymath of wide-ranging interests—philosophy, drama, and art as well as science—whose comments are perceptive and often witty. An example: In drama he admired Shakespeare and thought much less of his own countryman, Goethe. Why, inquired Lichtenberg, has Goethe come to be known as the German Shakespeare? He answered, "As the wood louse came by the name of millipede. Because no one could be bothered to count the legs."

Like Einstein, he was leery of the mystical. He wrote:

> As soon as someone begins to see everything in everything, his utterances usually become obscure; he begins to speak with the tongues of angels.

He filled many notebooks with aphorisms that reveal a mind that thought deeply, clearly, critically. Einstein—who also admired Shakespeare and thought much less of Goethe—relished Lichtenberg's terse comments. He called them "immortal fragments of thought." The older he gets, he observed in a letter from Princeton to his long-time friend Queen Elizabeth of Belgium, the more he appreciates his earlier colleague: "I know no one else who so plainly hears the grass grow." He offered to lend the Queen his volume of Lichtenberg's thoughts.

Some of the epigraphs concern God. Lichtenberg's culminating entry about God in his published notebooks was a question:

Is our conception of God anything more than per-
sonified incomprehensibility?

To read and think about his words at the
heads of the chapters is to join Albert Einstein, and
perhaps Queen Elizabeth of Belgium, in the savor-
ing of Lichtenberg's wit and not inconsiderable
wisdom.

Appendix 2

CONCERNING SOURCES

This information is for the reader who wishes to pursue further some of the specific material discussed. It supplements the Selected Bibliography.

Page vii

Einstein's words are from the preface he wrote to *Spinoza: Portrait of a Spiritual Hero*, Rudolf Kayser (New York: 1946). Kayser was Einstein's son-in-law.

Chapter 2

To learn about Minkowski and his ideas, see the article "Minkowski's Space-Time: From Visual Thinking to the Absolute World," Peter Louis Galison. In *Historical Studies in the Physical Sciences*, vol. 10 (Baltimore: 1979).

Philip Frank authoritatively discusses Einstein's life in Prague in *Einstein, His Life and Times* (New York: 1953). Frank assumed Einstein's position at the University of Prague when Einstein left.

Martin Buber tells of his conversation with Einstein in the essay, "Man and His Image-Work" in his book, *The Knowledge of Man: Selected Essays*, Maurice Friedman, ed. (New York: 1965).

For information about Ehrenfest, who became one of Einstein's better friends, see *Paul Ehrenfest, Vol. 1: The Making of a Theoretical Physicist*, Martin J. Klein (Amsterdam: 1970).

Chapter 3

Rudolf Carnap speaks of Einstein's attitude toward the "now" in *The Philosophy of Rudolf Carnap*, P. A. Schilpp, ed. (La Salle: 1963).

Masanao Toda considers the flow of time in his essay, "The Boundaries of the Notion of Time." In *The Study of Time III*, Fraser, Lawrence and Park, eds. (New York: 1978).

Chapter 4

Henry Pachter who, as a young boy, visited Einstein in Berlin, tells of Einstein's fondness for dirty jokes in *Weimar Etudes* (New York: 1982).

John Plesch, a medical doctor who became friendly with Einstein, writes of their relationship in his autobiography, *The Story of a Doctor* (London: 1947).

Einstein's questions to Blumenfeld are discussed in *Einstein: The Life and Times,* Ronald W. Clark (New York: 1971).

Esther Salaman reported Einstein's words to her in the article, "Remembering Einstein." In *Encounter* (London: 1979, April).

Chapter 5

Einstein's explanation of causality beginning "We must distinguish between causality as a postulate . . . " is quoted in *The Search for Meaning*, Alfred Stern (Memphis: 1971). Stern was professor of philosophy at the California Institute of Technology when Einstein was in residence as Visiting Professor. He reports these words from a conversation with Einstein in January, 1945. Stern states that Einstein reviewed and approved the material before it was published.

Einstein's explanation of his position on determinism from a historical standpoint is found in *Where Is Science Going?*, Max Planck (New York: 1932).

Chapter 6

Carnap tells of Einstein's statement on the future of physics in *The Philosophy of Rudolf Carnap. Ibid.*

Chapter 7

Muensterberg's comment, and Einstein's response to it, are contained in an interview with Einstein in *Glimpses of the Great*, George Sylvester Viereck (New York: 1930).

Chapter 8

The best source for information on Kurt Gödel is the mathematician Hao Wang, who had many philosophical discussions with him. See Wang's essay, "Gödel and Einstein as Companions." In *Doing Science: The Reality Club*, John Brockman, ed. (New York: 1991). See also Wang's book, *Reflections on Kurt Gödel* (Cambridge, Massachusetts: 1987).

Gödel's essay on time travel, and Einstein's response to it, can be found in *Albert Einstein: Philosopher-Scientist*, P. A. Schilpp, ed. (New York: 1951). His ideas on time travel are more fully expressed in *Kurt Gödel: Collected Works, Volume III* (Oxford: 1995).

Chapter 9

Some of Einstein's contemplative letters to Queen Elizabeth of Belgium can be found in *Einstein on Peace* (New York: 1968).

Chapter 10

William James' observations on the wider self are from the concluding chapter of his seminal volume, *The Varieties of Religious Experience* (New York: 1902).

Select Bibliography

Born, Max, editor. *The Born-Einstein Letters.* New York: Walker, 1971.

Broch, Hermann. *The Death of Virgil.* Oxford: Oxford University Press, 1983.

Bucky, Peter A. *The Private Albert Einstein.* Kansas City: Andrews and McMeel, 1992.

Curley, Edwin, editor. *The Collected Works of Spinoza.* Princeton: Princeton University Press, 1985.

Einstein, Albert. *About Zionism.* New York: Macmillan, 1931.

——. *Ideas and Opinions.* New York: Crown Publishers, 1954.

——. *Letters to Solovine.* New York: Philosophical Library, 1987.

Gandhi, Mohandas. *An Autobiography: The Story of My Experiments with Truth.* Boston: Beacon, 1957.

Lichtenberg, Georg Christoph. *Aphorisms.* London: Penguin, 1990.

Nathan, Otto and Norden, H., editors. *Einstein on Peace.* New York: Schocken, 1968.

Pais, Abraham. *'Subtle is the Lord . . . '*. New York: Oxford University Press, 1982.

Pedersen, Johannes. *Israel, Its Life and Culture.* London: Oxford University Press, vol. 1, 1926; vol. 2, 1940.

Samuel, Viscount. *Belief and Action.* London: Cassell, 1937.

Sayen, Jamie. *Einstein in America.* New York: Crown, 1985.

Schilpp, Paul Arthur, editor. *Albert Einstein: Philosopher-Scientist.* New York: Tudor, 1951.

Stern, J. P. *Lichtenberg: A Doctrine of Scattered Occasions.* Bloomington: Indiana University Press, 1959.

Wang, Hao. *Reflections on Kurt Gödel.* Cambridge: Massachusetts Institute of Technology Press, 1987.

Index

ABOUT THE AUTHOR

Robert N. Goldman is a pioneering computer scientist who has been granted over thirty patents both nationally and internationally. Companies in North America, Europe, and Japan—including IBM, NCR, and Fujitsu—have been licensees. His patents include one of the first computer software patents issued in the United States and one that led to the development of the magnetic-striped bank or credit card. For the past twenty years he also has been exploring Albert Einstein's philosophical and religious beliefs. He likes to point out that, although it is not generally known, Einstein also invented technical devices and was granted many patents in both Europe and the United States. Mr. Goldman lives with his wife, Judith, in Honolulu and London. They are the parents of two grown daughters.